Advances in Sustainability Science and Technology

The book series aims at bringing together valuable and novel scientific contributions that address the critical issues of renewable energy, sustainable building, sustainable manufacturing, and other sustainability science and technology topics that have an impact in this diverse and fast-changing research community in academia and industry.

The areas to be covered are

- Climate change and mitigation, atmospheric carbon reduction, global warming
- Sustainability science, sustainability technologies
- Sustainable building technologies
- Intelligent buildings
- Sustainable energy generation
- Combined heat and power and district heating systems
- Control and optimization of renewable energy systems
- Smart grids and micro grids, local energy markets
- Smart cities, smart buildings, smart districts, smart countryside
- Energy and environmental assessment in buildings and cities
- Sustainable design, innovation and services
- Sustainable manufacturing processes and technology
- Sustainable manufacturing systems and enterprises
- Decision support for sustainability
- Micro/nanomachining, microelectromechanical machines (MEMS)
- Sustainable transport, smart vehicles and smart roads
- Information technology and artificial intelligence applied to sustainability
- Big data and data analytics applied to sustainability
- Sustainable food production, sustainable horticulture and agriculture
- Sustainability of air, water and other natural resources
- Sustainability policy, shaping the future, the triple bottom line, the circular economy

High quality content is an essential feature for all book proposals accepted for the series. It is expected that editors of all accepted volumes will ensure that contributions are subjected to an appropriate level of reviewing process and adhere to KES quality principles.

The series will include monographs, edited volumes, and selected proceedings.

Tiago Miguel Ferreira
Editor

Multi-risk Interactions Towards Resilient and Sustainable Cities

Springer

Editor
Tiago Miguel Ferreira
School of Engineering, College of Arts,
Technology and Environment (CATE),
University of the West of England
(UWE Bristol)
Bristol, UK

ISSN 2662-6829 ISSN 2662-6837 (electronic)
Advances in Sustainability Science and Technology
ISBN 978-981-99-0744-1 ISBN 978-981-99-0745-8 (eBook)
https://doi.org/10.1007/978-981-99-0745-8

This Springer imprint is published by the registered company Springer Nature Singapore Pte Ltd.
The registered company address is: 152 Beach Road, #21-01/04 Gateway East, Singapore 189721, Singapore

Foreword

This book is dedicated to multi-hazard studies, including essentially seismic, flood, urban fire and landslides.

It starts with Chapter 1, introducing the topic of multi-hazard and their interactions towards a resilient and sustainable society, and its importance in urban planning and development.

Chapter 2 is dedicated to methods and tools for evaluating individual risks and managing them in the context of an urban area. Cascading effects are then introduced as multi-hazards of modern urban areas of complex interactions. The authors present two approaches to analyse the problem, the qualitative and semi-quantitative methodologies to represent risks, the first based on thresholds and the other on indexes. Climate change has been brought into the context of index terminology. The authors prefer the second one for reasons of better population perceptibility. Then, they introduce the scale of analyses, one of the most important parameters when we deal with the geography territory.

Chapter 3 is dedicated to the analysis of what the authors call social vulnerability using a large number of social indexes. Illustrations of the developments in this chapter and all the others are made through an application to the Lisbon Metropolitan Area (LMA). But the concepts could be used in any other territory with similar constraints.

Chapter 4 is dedicated to flood risk assessment in the LMA, Chapter 5 is to fire susceptibility and Chapter 6 is to seismic vulnerability (excluding tsunami, which will be dealt with in Chapter 8) and risk assessment.

Chapter 7 analyses landslide hazards by the Analytic Hierarchy Process (AHP) methodology.

Final Chapter 8 is tentative to consider the multi-hazard susceptibility assessment in the land use planning of LMA, also considering the tsunami, coastal erosion and cliff retreat in a climate change environment.

The book is intended to fulfil various audacious objectives considered individual risks and in a global multi-hazard panorama. It pretends to be a helpful text for scientists not familiar with other topics of the multi-hazard collection besides their own and a practical guideline for end-users and stakeholders, mainly practitioners and decision-makers.

The book is well-balanced among the topics above, and the new jargon is a good complement for some readers and an introduction to others.

The application to LMA is an excellent choice as we are in a larger urban area with a mix of the highly concentrated population and the less populated area, zones with an ancient patrimony and new moderate size, 10-15 stories high modern buildings. Social vulnerability is a topic of great importance that governs most possible hazards and significant risks and impacts.

The unit of work changes from hazard to hazard, being the county in some cases or parishes in others, or at the urban block/susceptibility area level as in a detailed analysis of landslides, reflecting the authors´ experience. While Census block information is the primary data source to analyse social vulnerability, urban fires are studied at the parish level. Urban types of equipment are based on a radius of influence used to define criticality through principal component analyses (PCA). Finally, aggregation on counties is made.

Census data may not be the most qualified source of information because the parameter values depend very much on the preparation and prior knowledge of the agents doing it, and sometimes apparent errors are committed, especially if we talk of a large scale of study. Most probably, we can use Artificial Intelligence/Machine Learning (AI/ML) shortly to help in decreasing uncertainty on census data gathering. As an example, we can say that AI/ML can help in identifying in a building the number of stories, the epoch of construction/rehabilitation, the state of conservation, etc. speeding the data-acquiring system and reducing budgets.

In addition, the quality of the final results depends on the detailing techniques used in the individual studies. There are cases of simulation where we can validate the models using past events, which should be exercised whenever possible. Still, there are other cases where it is not easy to calibrate the models.

A question that arises in many texts dealing with various types of hazards is the meaning of terminology/glossary/taxonomy, which has not yet been accepted worldwide. On the contrary, we keep seeing different terms to designate the same phenomenon, and some confusion may arise among specialists in a specific hazard. Lately, with the introduction of new topics about various hazards, the situation may be worse. The Glossary of UNESCO serves this moment to give a minimum understanding of different hazards and risks. Still, in the future, we badly need a better and more complete series of terms to attain more consistency. Specialized terminology is already available in several areas, such as the case of the Global Earthquake Model (GEM), which is very much oriented to earthquake engineering, whereas the United States Geological Survey (USGS) goes to geological Hazards.

An extensive collection of papers organized by chapters (40 on average with repetitions) completes the text. Each chapter is a text that can stand by itself, useful for individual interest reading. So that if someone only wants to look into a single topic, this chapter does not require the reading of other chapters. This option, made by a great team of experts in the territory's geography, creates repetitions on the main description of LMA.

I want to call attention to the problem of multi-hazard that LMA already suffered during the 1755 Lisbon earthquake, which was probably one of the first areas struck by a series of multi-hazard events that started with a series of significant intensity and long duration ground shakings causing the failure of a large number of collapses. A strong tsunami followed within tens of minutes and devastated a large area downtown. The tragedy continued with an urban fire that was difficult to extinguish due to the precarious damage and lack of means to fight. More hazards can be associated with events like this, such as the liquefaction of particular soils and even the existence of landslides in the hills of greater inclination. Overall, the 1755 event caused the LMA an estimated number of victims in the order of 20 000 to 40 000 persons out of a population of 150 000-200 000 and caused significant impacts in Spain and Morocco. For Portugal, the estimate is 50 to 100% of the GDP at the epoch.

In terms of the impact on the constructed park, 30% collapsed, 30% was inhabitable, 30% could be used with great caution, and only 10% was in conditions of being occupied.

Carlos Sousa Oliveira
Professor Emeritus
Instituto Superior Técnico
Lisbon, Portugal

About This Book

Over the past years, a considerable volume of work has been conducted to assess urban risks resulting from individual natural hazards. However, assessing the probable cumulative impacts of multiple hazards has not yet entered the mainstream of research and urban management practice, which represents a significant drawback in the current climate change context. This book aims to contribute to tackling this issue by offering a comprehensive discussion of the various stages involved in identifying, assessing, and managing multi-hazard risk in urban areas—covering from the identification and characterisation of the exposed elements to the analysis of the impact of individual hazards (including earthquake, flood, fire, and landslide) and the assessment and management of risk posed by multiple hazards in a climate change context. After being presented in a clear and structured way, all the concepts and approaches addressed in the book are applied to the Lisbon Metropolitan Area, one of the most vibrant and dynamic metropolitan areas of the globe, allowing for an applied understanding of the covered topics. Besides offering solid theoretical grounding, this book provides practical guidelines on how the outputs coming from vulnerability and risk assessment approaches can be used to outline effective risk mitigation and emergency planning strategies, intending to be both a reference book and a practical guideline for end-users, including research scientists, practitioners, and decision-makers.

Contents

Editor and Contributors

About the Editor

Dr. Tiago Miguel Ferreira is a Lecturer in Civil Engineering at the School of Engineering of the University of the West of England (UWE Bristol), United Kingdom, and an invited Assistant Professor at the University of Coimbra, Portugal. Dr. Ferreira's research focuses on the structural vulnerability of historical buildings and urban areas to natural and anthropogenic hazards, specifically earthquakes, fires, floods, and landslides. More recently, he has expanded his focus to include the interaction between different hazards and both physical and social vulnerability, both in the context of single, compound and cascading hazards. Recognised as among the 2% top-cited scientists in the world by Elsevier BV and Stanford University (2021 and 2022) twice, Dr. Ferreira is a highly accomplished academic in his fields of expertise. He has co-authored nearly 200 scientific and technical publications, including dozens of research articles in some of the most reputed international journals. He has also edited several books on the topics of vulnerability and risk assessment and participated in and coordinated many research projects in these fields. Currently, Dr. Ferreira is a co-Editor-in-Chief of 'GeoHazards', a multidisciplinary journal devoted to theoretical and applied research across the whole spectrum of geomorphological hazards, and a Section Editor-in-Chief of 'Fire', a wide-spectrum journal about the science, policy, and technology of fires and how they interact with communities and the environment.

Contributors

Ana Cardoso Centre for Geographical Studies, Institute of Geography and Spatial Planning (IGOT), LA TERRA, University of Lisbon, Lisbon, Portugal

José Carlos Domingues Institute for Sustainability and Innovation in Structural Engineering (ISISE), Department of Civil Engineering, University of Minho, Guimarães, Portugal

Tiago Miguel Ferreira School of Engineering, College of Arts, Technology and Environment (CATE), University of the West of England (UWE Bristol), Bristol, UK

Ricardo A. C. Garcia Centre for Geographical Studies, Institute of Geography and Spatial Planning (IGOT), LA TERRA, University of Lisbon, Lisbon, Portugal

Paulo B. Lourenço Institute for Sustainability and Innovation in Structural Engineering (ISISE), Department of Civil Engineering, University of Minho, Guimarães, Portugal

Raquel Melo Institute of Earth Sciences, School of Science and Technology, University of Évora, Évora, Portugal

Sérgio Cruz Oliveira Centre for Geographical Studies, Institute of Geography and Spatial Planning (IGOT), LA TERRA, University of Lisbon, Lisbon, Portugal

Carolina Pais Centre for Geographical Studies, Institute of Geography and Spatial Planning (IGOT), LA TERRA, University of Lisbon, Lisbon, Portugal

Susana Pereira Faculty of Arts and Humanities, Geography Department, University of Porto, Porto, Portugal;
Centre for Geographical Studies, Institute of Geography and Spatial Planning (IGOT), LA TERRA, University of Lisbon, Lisbon, Portugal

Eusébio Reis Centre for Geographical Studies, Institute of Geography and Spatial Planning (IGOT), LA TERRA, University of Lisbon, Lisbon, Portugal

Pedro Pinto Santos Centre for Geographical Studies, Institute of Geography and Spatial Planning (IGOT), LA TERRA, University of Lisbon, Lisbon, Portugal

Ângela Santos Centre for Geographical Studies, Institute of Geography and Spatial Planning (IGOT), LA TERRA, University of Lisbon, Lisbon, Portugal

Teresa Vaz Centre for Geographical Studies, Institute of Geography and Spatial Planning (IGOT), LA TERRA, University of Lisbon, Lisbon, Portugal

Maria Xofi Faculty of Engineering Technology, University of Twente, Enschede, The Netherlands;
Institute for Sustainability and Innovation in Structural Engineering (ISISE), Department of Civil Engineering, University of Minho, Guimarães, Portugal

José Luís Zêzere Centre for Geographical Studies, Institute of Geography and Spatial Planning (IGOT), LA TERRA, University of Lisbon, Lisbon, Portugal

Chapter 1
An Introduction to Multi-hazard Risk Interactions Towards Resilient and Sustainable Cities

Tiago Miguel Ferreira and Pedro Pinto Santos

Abstract The relationship between disaster resilience and sustainability in the context of urban risk has gained significant attention in recent years as the research and technical community work towards a safer, more sustainable way of living. Urban risk is a complex matrix that involves multiple elements at risk, hazards, temporal scales, and vulnerabilities, and this is why traditional risk assessment approaches that focus on addressing the impacts of a single hazard are inadequate for effectively assessing and managing urban risk, particularly in the current climate change context. With this in mind, the present chapter provides an introduction to the concept of multi-hazard risk and its relevance to resilient and sustainable cities by listing and briefly discussing the types of natural hazards that impact cities the most and examining the importance of risk assessment and management in reducing the risks posed by these hazards. The chapter also explores strategies for building resilience in cities, including the strengthening of physical infrastructure and the enhancement of social and economic resilience, and concludes by discussing future directions for research and practice in multi-hazard risk management for resilient and sustainable cities.

Keywords Multi-hazard risk · Risk assessment · Risk management · Resilience · Sustainable cities

T. M. Ferreira (✉)
School of Engineering, College of Arts, Technology and Environment (CATE), University of the West of England (UWE Bristol), Bristol, UK
e-mail: Tiago.Ferreira@uwe.ac.uk

P. P. Santos
Centre for Geographical Studies, Institute of Geography and Spatial Planning (IGOT), LA TERRA, University of Lisbon, Lisbon, Portugal
e-mail: pmpsantos@campus.ul.pt

T. M. Ferreira (ed.), *Multi-risk Interactions Towards Resilient and Sustainable Cities*, Advances in Sustainability Science and Technology,
https://doi.org/10.1007/978-981-99-0745-8_1

1 Introduction

Since the global framework for disaster risk reduction [44] was defined and approved in Sendai, Japan, there has been a greater focus on linking disaster resilience and sustainability through policies and tools that support risk-informed sustainable development. The relationship between the concepts of disaster resilience and sustainability in the context of urban risk has also been developing, and their role towards a safer, more sustainable way of living is now understood in a clear and more comprehensive way by the research and technical community working in this field [11].

Urban risk is a multi-dimensional matrix that often includes multiple elements at risk, such as people, buildings, and infrastructures, multiple hazards, multiple temporal scales, as well as multiple types of vulnerabilities. For this reason, unlike current practice, addressing the impacts of one single hazard is not sufficient when assessing and managing urban risk, and it is, therefore, necessary to replace current single-hazard approaches with multi-hazard risk assessment.

Multi-hazard risk refers to the potential impacts and consequences of multiple hazards on a particular location or system [24]. These hazards can be natural, human-induced, or socioeconomic in nature and may include events such as earthquakes, floods, landslides, fires, industrial accidents, terrorism, financial crises, and pandemics. Understanding and managing multi-hazard risk is critical for building resilient and sustainable cities, as it enables planners and policymakers to identify and assess potential hazards and vulnerabilities, develop risk management strategies, and implement measures to reduce the impacts of hazards on communities and infrastructure. In this chapter, we will be focusing particularly on natural hazards, although some of the considerations provided here also apply to other types of hazards.

The importance of considering multi-hazard risk in urban planning and development cannot be overstated. As extensively discussed in literature—see, for example, Lau et al. [25] or Pelling [33]—urbanization and population growth have led to the concentration of people and assets in cities, making them more vulnerable to the impacts of hazards. At the same time, cities also provide important economic, social, and cultural benefits, and are key drivers of global development and progress. Ensuring the resilience of cities is therefore essential for protecting communities, preserving economic, social and cultural assets [38], and promoting sustainable development.

This chapter aims to provide an introduction to the concept of multi-hazard risk and its relevance to resilient and sustainable cities—the topic of the project "MIT-RSC—Multi-risk Interactions Towards Resilient and Sustainable Cities", funded by the Portuguese Foundation for Science and Technology (FCT) under the MIT Portugal Program at the 2019 PT call for Exploratory Proposals in "Sustainable Cities", which was the basis for this book. The chapter begins by discussing the types of hazards that can impact cities and then discusses risk assessment and management strategies for reducing the impacts of these hazards. Finally, it concludes by exploring strategies for building resilience in cities, and identifies key directions for future research and practice in this area.

2 Natural Hazards in Urban Settings

Cities are exposed to a variety of hazards that can have significant impacts on communities and infrastructure. As mentioned earlier, these hazards can be natural, human-induced, or socioeconomic in nature, and can include very different types of events. Understanding the types of hazards that cities are exposed to is a critical first step in developing effective risk assessment and management strategies. Understanding the types of hazards that cities are exposed to is crucial for developing risk assessment and management strategies that effectively reduce their impacts. This is because specific hazards have different characteristics and impacts that need to be considered in the risk assessment process. As discussed in detail by Dickson et al. [10], tailored risk management strategies can then be developed based on the hazards that the city is exposed to. For example, a city exposed to earthquakes may need to focus on strengthening buildings and infrastructure, while a city exposed to floods may need to improve drainage and flood protection systems. This targeted approach helps to ensure that resources are used effectively and efficiently, as risk management measures will be focused on the hazards that pose the greatest risk to the city.

According to the Federal Emergency Management Agency [12], "*natural hazards are defined as environmental phenomena that have the potential to impact societies and the human environment*". These hazards, which are being increasingly influenced by climate change, can have significant impacts on cities, including physical damage to buildings and infrastructure, loss of life, and disruption of essential services. As previously mentioned, understanding the types of natural hazards that cities may be exposed to is essential for designing efficient risk assessment and management strategies. That is why it is probably worth including at this point a brief definition of the hazards that most commonly affect urban settings. It is also worth noting that the hazards listed below are the ones that will be addressed and discussed in greater detail in the following chapters of this book.

2.1 Earthquakes

Earthquakes are sudden, violent shaking of the ground caused by the movement of tectonic plates in the Earth's crust. They can cause damage to buildings and infrastructure, particularly in areas with a high concentration of unreinforced masonry or poorly constructed buildings (Fig. 1). Earthquakes can also trigger landslides, which can further damage buildings and infrastructure and block roads, disrupting transportation.

Fig. 1 Examples of urban masonry buildings damaged by earthquakes

Robert Mallet and John Milne were pioneers in studying how and why buildings were damaged during earthquakes. According to Cousins and Smith [7], on 13 October 1900, John Milne (1850–1913) wrote regarding his 1892 book on earthquakes: "*if you compare the contents of this volume with its reproduction, and a companion volume called 'Seismology' issued in 1898, you will realize the rate at which a neglected study is advancing*." In his observations, Milne noted that some societies had developed solutions for resisting earthquake damage and incorporated them into local construction practices, while in others, architects and engineers were unaware of the need to design for future shaking, a phenomenon that can be partially explained by the frequency of earthquakes in certain regions and the corresponding risk perception and seismic culture of the populations in those regions [27]. In his catalogue of destructive earthquakes [31], Milne estimated that over 12 million people had been killed by earthquakes in the 2,000 years prior to 1900. As noted by Bilham [7], he would have been astonished to see that, even with the advances in earthquake engineering that occurred in the century following his words, earthquakes would claim an additional 2 million lives.

2.2 *Floods*

Floods are the most frequent type of natural disaster and occur when an overflow of water submerges land that is usually dry. They can be caused by heavy rainfall, rapid snowmelt, or a storm surge from a tropical cyclone or tsunami in coastal areas. Depending on their extension and magnitude, floods can cause widespread devastation, resulting in loss of life and damage to personal property.

According to the United Nations Office for Disaster Risk Reduction [43], floods have affected more than 2 billion people worldwide between 1998 and 2017 alone (Fig. 2). People who live in floodplains or non-resistant buildings or lack warning systems and awareness of flooding hazards are particularly vulnerable to floods.

Fig. 2 Floods in Bingley, England, in 2015 (left) and in Venice, Italy, in 2018 (right)

Floods can also disrupt essential services, such as public health infrastructure, transportation, communication, and power, which can be particularly meaningful in urban areas, as discussed in great detail by Jha et al. [22] or Hammond et al. [19].

2.3 Landslides

Landslides, or mass movements, are the movement of rock, soil, or debris down a slope. As noticed by Alexander [2] in his pioneering article on urban landslides, landslides are often triggered by other hazards, such as earthquakes, storms, or heavy rainfall, and per se, rarely cause disasters that attract international attention. However, in many countries, smaller-scale landslide disasters, which involve damage valued at hundreds of thousands of dollars and may cause a few fatalities, are both numerous and frequent [2].

It is plausible to believe that the occurrence and impact of urban landslides will tend to increase in future due to two main reasons. Firstly, as a result of the pressures created by population growth—a topic that has already been widely treated in literature applied to different types of hazards; see, for example, Radeloff et al. [20], Hemmati et al. [21], He et al. [35]. Secondly, people are attracted to building on hillsides due to their natural beauty. In Olshansky's words [32], "*Hillsides pose unique problems for the construction and maintenance of human settlements. They are prone to natural hazards, and they topographically constrain the design of settlements. For these reasons, hillside lands often remain vacant long after adjacent valley floors are urbanized. Despite the constraints, they are attractive places to live because of the views and because of the sense of being close to nature.*" Thus, much urban expansion is expected to take place in hillside areas, and consequently, ground failure by landsliding will likely be one of the most significant geological hazards affecting urban settings in future [37].

2.4 Urban Fires

Urban fires are a significant concern in cities around the world. The already mentioned high concentration of buildings, people, and resources in urban areas makes them particularly vulnerable to the impacts of fires. Urban fires can have a range of causes, including electrical malfunctions, accidents resulting from human activities such as cooking, smoking, using open flames, candle usage, or arson. Urban fires can also be triggered by natural hazards, such as earthquakes. Post-earthquake fires are a low-probability phenomenon but can have high consequences in terms of human loss and severe damage to property [23]. According to Cousins and Smith [8], fire losses are usually zero for most earthquakes. However, when a fire does occur after an earthquake, the losses can surpass those caused by the earthquake itself. The San Francisco earthquake in 1906 is a prime example of this—over 80% of the reported damages were due to fire, and more than 28,000 buildings were destroyed [36].

However, the consequences of urban fires per se can be devastating. They can cause significant damage to buildings and infrastructure and lead to injuries and fatalities. According to the World Bank [1], more than 180,000 people die each year in fires or from burn-related injuries worldwide, with 95% of those deaths occurring in low- and middle-income countries where risks rise proportionally with rapid urbanization (Fig. 3). Inadequate urban planning, infrastructure, and construction practices related to fire are among the top factors that contribute most to the potential for conflagration, ignition, and the spread of fires in buildings and urban spaces [13]. In addition, urban fires can have significant environmental impacts, as they release pollutants into the air and can alter local ecosystems.

Effective prevention and management of urban fires is critical to minimizing their impacts. This can include measures such as building codes and regulations that aim to reduce the risk of fires through the implementation of active and passive fire mitigation measures, as well as fire prevention education and outreach programs.

Fig. 3 Urban fires in Leeds, England (left), and in Hargeisa, Somalia (right), both in 2022

3 Multi-hazard Risk Interactions

As mentioned in the introductory section of this chapter, urban risk is a multifaceted concept that involves multiple elements at risk, multiple temporal scales, multiple vulnerabilities, and multiple hazards. This means that cities are often exposed to multiple hazards at the same time, and these hazards can interact in complex ways to create new risks or amplify existing ones. These multifaceted risk interactions can be difficult to predict and can have significant impacts on communities and infrastructure. Therefore, understanding the ways in which hazards can interact and the risks that these interactions create is crucial in developing effective risk assessment and management strategies.

As awareness of the complexities associated with multifaceted risk interactions has grown, a wide variety of terms have emerged to describe them [9, 46]. In an effort to clarify this diversity, Tilloy et al. [42] have reviewed and categorized the various terms into five main types of hazards:

1. Independent: As the name suggests, independent hazard events are events that overlap in space and time but are unrelated to one another. An example of this might be the occurrence of wildfires during the COVID-19 pandemic—they were coincident in spatial location and time, but were completely unrelated. Some authors also include in this category cases where two unrelated hazards impact the same area, at different times (e.g., a flood occurring a few weeks after an urban fire, or an earthquake after a cyclone).
2. Triggering (cascading): Triggering or cascading events occur when one hazard triggers another hazard, creating a chain reaction of events, meaning that triggering events always imply a primary and a secondary hazard. According to [17], any natural hazard might trigger none, one or more than one secondary hazard. For example, an earthquake can trigger a landslide, which could then block a river and cause a flood. The secondary hazard can be the same as or different from the primary hazard [42]. Cascading events can have significant impacts on cities, as each hazard in the chain reaction can cause damage and disruption on its own, and the combined effects of all of the hazards can be much greater than the sum of their individual impacts (e.g., post-earthquake fires, as discussed in Sect. 2.4).
3. Change conditions: Change conditions occur when a primary hazard changes the probability of a second hazard occurring by altering environmental conditions. For example, the impact of wildfires on the risk of landslides or flooding events is a good example of this phenomenon.
4. Compound (association): Compounding hazards involve the simultaneous occurrence of different hazards that are dependent on one another and, when combined, create new or amplify existing risks. In this case, there is no primary and secondary hazard as the different hazards occur simultaneously [42]. For example, a hurricane could produce heavy rainfall that causes a flood, which, in turn, could damage buildings and infrastructure, making them more vulnerable to future hazards. The compound effect of these hazards can have significant

impacts on cities, as the risks resulting from the combination of the hazards can be much greater than the individual risks of each hazard, like in the case of cascading events.

5. Mutual exclusion: Two natural hazards are said to be mutually exclusive if they have a negative dependence on one another. For example, heavy rain may help to extinguish a wildfire. As noticed by Tilloy et al. [42], since a negative dependence of two hazards does not lead to increased impact (which would be a positive dependence), there is less literature on this type of hazard interaction.

From the set of examples provided above, it is clear that understanding the ways in which hazards can interact, and the risks that these interactions create is a crucial step in developing effective risk assessment and management strategies that can reduce the impacts of multiple hazards (as will be discussed in the next section). By considering the potential interactions between hazards, cities can develop risk management measures that address the combined risks of multiple hazards, rather than just the individual risks of each hazard.

4 Risk Assessment and Management

Risk assessment is the process of identifying, analyzing, and evaluating the risks to which a city is exposed in order to understand the likelihood and potential impacts of these risks. On the other hand, risk management involves developing and implementing strategies to reduce these risks in order to minimize the likelihood and potential impacts. It is essential that risk management be based on a comprehensive and accurate understanding of the risks to which cities are exposed (obtained in the risk assessment stage). Otherwise, the efficiency of the strategies implemented during the risk management stage may be compromised. Only by ensuring that the risk is comprehensively and accurately assessed and that the results of this assessment are processed, integrated, and considered in the risk management stage can we develop and implement strategies that can efficiently reduce the risks to which a city is exposed, contributing to more resilient and sustainable urban areas.

Risk assessment and management strategies for cities should consider the potential interactions between hazards as well as the individual risks of each hazard. By considering the potential interactions between hazards, decision-makers can develop risk management measures that address the combined risks of multiple hazards, rather than just the individual risks of each hazard. As discussed in Sect. 3, the combined risks of these hazards can often be much greater than the individual risks of each hazard, so it is key to accurately assess the risks resulting from multiple hazards in areas that are potentially exposed to multiple hazard events.

There are many different approaches to assessing and managing risk in urban settings. However, in general, urban risk assessment is usually described as a four-step process involving: the identification and characterization of hazards that can

Fig. 4 Four steps of risk assessment

potentially affect the city, the assessment of the exposed elements, and the evaluation of the vulnerability of those elements—as illustrated in Fig. 4.

Hazards are typically represented on maps that show the locations and intensities of all the hazards that can potentially affect the city. These maps serve a purpose beyond simply representation, as they are key tools for understanding the risks that these hazards pose to a city. They are fundamental, for example, for identifying the exposed elements in the second step of the risk assessment process, see Fig. 4. When the elements exposed to the hazards considered in the assessment are identified, it is necessary to assess their vulnerability, i.e., to identify the factors that make these elements vulnerable to those hazards, such as the age and condition of buildings and infrastructure, and the demographics and socioeconomic status of the population. It is worth noting that exposure must be disaggregated by individual hazard, as vulnerabilities are hazard-specific, i.e., a specific building or infrastructure can be vulnerable to an earthquake but not to a flood.

Finally, the risk assessment stage involves analyzing the likelihood and potential impacts of the hazards, based on the magnitude and extent of the identified hazards and the vulnerability of the exposed elements to these specific hazards, in order to understand the risks to which a city is exposed.

Once the risks to which a city is exposed have been identified and understood, risk management strategies can be developed and implemented in order to reduce these risks. These strategies can include measures such as strengthening buildings and infrastructure to make them less vulnerable to hazards [15, 28], developing evacuation and emergency response plans [5, 45], and implementing land use and zoning regulations. Risk management strategies should also consider the potential impacts of climate change, which can drastically increase the extent, magnitude, and types of hazards to which cities are exposed. This is particularly important for coastal cities, which are especially threatened by the impacts of sea level rise.

5 Creating More Resilient Urban Areas

The popularity of the word "resilience" has increased significantly in both academic and policy discourse in recent years, with numerous explanations for this dramatic rise [29]. Probably one major reason for this is that resilience theory provides insights into complex socio-ecological systems and their sustainable management, particularly with respect to climate change, see, for example, Leichenko [26], Solecki et al. [34], Pierce et al. [41]. In particular, resilience has become an attractive perspective with respect to cities, often theorized as highly complex, adaptive systems [4, 18], as noted by Meerow et al. [29], who define urban resilience as "*the ability of an urban system-and all its constituent socio-ecological and socio-technical networks across temporal and spatial scales-to maintain or rapidly return to desired functions in the face of a disturbance, to adapt to change, and to quickly transform systems that limit current or future adaptive capacity.*"

In the context of the present review, this complex definition of urban resilience can probably be simplified to "the ability of a city to withstand, recover from, and adapt to the impacts of natural hazards." Regardless of the definition we choose to use, it is clear that enhancing the resilience of urban areas is a critical step in reducing the risks posed by natural hazards and creating more sustainable and livable communities.

There are many different strategies that cities can use to build urban resilience, and the most appropriate strategies will depend on the specific hazards that a city is exposed to, as well as its resources and capabilities. One key strategy for building resilience in cities is to strengthen the physical infrastructure of the city, such as buildings, roads, bridges, and utilities—which, in other words, means making it less vulnerable. As mentioned before, this can involve retrofitting existing infrastructure to make it more resistant to hazards, or building new infrastructure to higher standards of performance. For example, buildings can be designed and constructed to be more resistant to earthquakes and storms using isolation [40], structural reinforcing [6, 14, 16, 30], and wind-resistant design [39]. Infrastructure can also be designed and constructed to be more resilient to floods, using techniques such as elevation, flood walls, and flood gates [3].

In addition to strengthening the physical infrastructure of a city, it is also crucial to consider the social and economic resilience of the city. Building the capacity of communities to withstand and recover from the impacts of hazards is a critical aspect of social resilience. This can involve building the capacity of communities to respond to disasters, such as through training and education programs, or building the capacity of communities to adapt to changing hazards, such as through economic diversification and diversification of livelihoods. For instance, communities can be trained in disaster preparedness and response techniques, such as first aid, search and rescue, and emergency evacuation, in order to better protect lives and property during a disaster. Communities can also be supported in diversifying their livelihoods and economic activities in order to reduce their dependence on a single industry or sector that may be vulnerable to hazards, such as agriculture or tourism.

Enhancing the economic resilience of a city is also critically important. This can involve, for example, supporting the development of a diverse and robust economy with a mix of industries and sectors that are less vulnerable to hazards. It can also involve supporting the development of small and medium enterprises, which are often more resilient to hazards than larger enterprises, as they are more flexible and adaptable. In addition, it can involve supporting the development of a strong and well-functioning financial system, with access to credit and insurance, in order to enable businesses and households to recover from the impacts of hazards. Some of these aspects will be treated in more detail in the next chapters of this book.

6 Future Directions for Research and Practice in Multi-hazard Risk for Resilient and Sustainable Cities

As the impacts of climate change and other natural hazards continue to increase, the need for effective multi-hazard risk management for resilient and sustainable cities becomes more pressing. There are many areas where further research and practice are needed in order to better understand and address the risks that cities are exposed to. Some key directions for future research and practice in this field include the following:

1. Improved understanding of multi-hazard risk interactions: In order to develop effective risk management strategies for cities, it is critical to better understand the ways in which different hazards interact with one another, as well as the combined impacts of these hazards. Further research is needed to identify the key factors that contribute to multi-hazard risk interactions, and to develop methods for quantifying and predicting these interactions.
2. Development of more comprehensive risk assessment and management approaches: Current risk assessment and management approaches tend to focus on individual hazards, rather than on the combined risks of multiple hazards. In order to more effectively reduce the risks posed by multi-hazard risks, it is necessary to develop more comprehensive approaches that consider the interdependent nature of hazards and the combined impacts of these hazards. This may involve developing new methods for analyzing and quantifying multi-hazard risks or adapting existing methods to better account for multi-hazard risks.
3. Integration of risk assessment and management into planning and decision-making processes: In order to effectively reduce the risks posed by natural hazards, it is important to integrate risk assessment and management considerations into the planning and decision-making processes of cities. This may involve the development of new planning and decision-making tools and methodologies that explicitly consider risk or the integration of risk considerations into existing planning and decision-making processes.
4. Improved communication and engagement with stakeholders: Effective risk management for cities requires the participation and engagement of a wide range of stakeholders, including government officials, community leaders, businesses, and the general public. In order to effectively engage these stakeholders, it is important to develop improved communication and engagement strategies that effectively convey the risks that cities are exposed to, and the steps that can be taken to reduce these risks.
5. Research on the effectiveness of risk management strategies: In order to better understand which risk management strategies are most effective in reducing the risks posed by natural hazards, it is important to conduct research on the effectiveness of these strategies. This may involve evaluating the effectiveness of different risk management approaches in different contexts or comparing the effectiveness of different risk management strategies for specific hazards. This

research can help to inform the development of more effective risk management strategies for cities, and can help to identify best practices for risk management.

By focusing on these key areas, researchers and practitioners can help to create more resilient and sustainable cities and contribute to the development of a more secure and sustainable future for our communities.

7 Final Remarks

In this chapter, we have examined the concept of multi-hazard risk and its importance in urban planning and development. We have also listed and briefly reviewed some of the natural hazards that affect cities the most and discussed the importance of risk assessment and management in reducing the risks posed by these hazards. Finally, we have discussed a few strategies for building resilience in cities and pointed out some future directions for research and practice in multi-hazard risk management for resilient and sustainable cities.

Through this discussion, we have seen that multi-hazard risk is a complex and multifaceted issue, and that effective risk management requires a comprehensive approach that considers the interdependent nature of hazards and the multiple impacts of these hazards. After briefly discussing the concept of "urban resilience", we have also seen that enhancing urban resilience is a crucial step in reducing the risks posed by natural hazards and creating more sustainable and livable communities.

As the impacts of climate change and other natural hazards continue to increase, it is more important than ever to better understand and address the risks that cities are exposed to. We hope that the contents of this book will help researchers, practitioners, and decision-makers to better understand this complex topic and, in this way, contribute to the development of more resilient, secure, and sustainable cities and communities.

Acknowledgements The project "MIT-RSC—Multi-risk Interactions Towards Resilient and Sustainable Cities" (MIT-EXPL/CS/0018/2019) leading to this work is co-financed by the ERDF—European Regional Development Fund through the Operational Program for Competitiveness and Internationalization—COMPETE 2020, the North Portugal Regional Operational Program—NORTE 2020 and by the Portuguese Foundation for Science and Technology—GCT under the MIT Portugal Program at the 2019 PT call for Exploratory Proposals in "Sustainable Cities". Pedro P. Santos is financed through FCT I.P., under the contract CEECIND/00268/2017.

References

1. Abrassart T, Meacham B, Sakoda K, Moullier T (2021) From urban fire risk to effective mitigation–building safer & healthier cities. In: Sustainable cities: building inclusive, resilient, and sustainable communities. The World Bank Group, Washington DC, USA

2. Alexander D (1989) Urban landslides. Prog Phys Geogr Earth Environ 13:157–189. https://doi.org/10.1177/030913338901300201
3. Barsley E (2020) Retrofitting for flood resilience: a guide to building & community design. Routledge
4. Batty M (2008) The size, scale, and shape of cities. Science (80)319:769–771
5. Bernardini G, Ferreira TM (2021) Combining structural and non-structural risk-reduction measures to improve evacuation safety in historical built environments. Int J Archit Herit, 1–19. https://doi.org/10.1080/15583058.2021.2001117
6. Bhattacharya S, Nayak S, Dutta SC (2014) A critical review of retrofitting methods for unreinforced masonry structures. Int J Disaster Risk Reduct 7:51–67. https://doi.org/10.1016/j.ijdrr.2013.12.004
7. Bilham R (2009) The seismic future of cities. Bull Earthq Eng 7:839–887. https://doi.org/10.1007/s10518-009-9147-0
8. Cousins WJ, Smith WD (2004) Estimated losses due to post-earthquake fire in three New Zealand cities. In: Proceedings of new zealand society of earthquake engineering conference, Christchurch, New Zealand
9. Cutter SL (2018) Compound, cascading, or complex disasters: what's in a name? Environ Sci Policy Sustain Dev 60:16–25. https://doi.org/10.1080/00139157.2018.1517518
10. Hoornweg DJL, Asmita D, Tiwari EB (2012) Urban risk assessments. The World Bank
11. Elmqvist T, Andersson E, Frantzeskaki N et al (2019) Sustainability and resilience for transformation in the urban century. Nat Sustain 2:267–273. https://doi.org/10.1038/s41893-019-0250-1
12. FEMA (2021) National risk index: technical documentation. Washington DC, USA
13. Ferreira TM (2022) Fire risk assessment and safety management in buildings and urban spaces-a new section of *Fire* journal. Fire 5(3):74. https://doi.org/10.3390/fire5030074
14. Ferreira TM, Costa AA, Arêde A et al (2016) In situ out-of-plane cyclic testing of original and strengthened traditional stone masonry walls using airbags. J Earthq Eng 20:749–772. https://doi.org/10.1080/13632469.2015.1107662
15. Ferreira TM, Maio R, Vicente R (2017) Analysis of the impact of large scale seismic retrofitting strategies through the application of a vulnerability-based approach on traditional masonry buildings. Earthq Eng Eng Vib 16:329–348. https://doi.org/10.1007/s11803-017-0385-x
16. Ferreira TM, Mendes N, Silva R (2019) Multiscale seismic vulnerability assessment and retrofit of existing masonry buildings. Buildings 9:91. https://doi.org/10.3390/buildings9040091
17. Gill JC, Malamud BD (2014) Reviewing and visualizing the interactions of natural hazards. Rev Geophys 52:680–722. https://doi.org/10.1002/2013RG000445
18. Godschalk DR (2003) Urban hazard mitigation: creating resilient cities. Nat Hazards Rev 4:136–143. https://doi.org/10.1061/(ASCE)1527-6988(2003)4:3(136)
19. Hammond MJ, Chen AS, Djordjević S et al (2015) Urban flood impact assessment: a state-of-the-art review. Urban Water J 12:14–29. https://doi.org/10.1080/1573062X.2013.857421
20. He C, Huang Q, Bai X et al (2021) A global analysis of the relationship between urbanization and fatalities in earthquake-prone areas. Int J Disaster Risk Sci 12:805–820. https://doi.org/10.1007/s13753-021-00385-z
21. Hemmati M, Ellingwood BR, Mahmoud HN (2020) The role of urban growth in resilience of communities under flood risk. Earth's Futur 8https://doi.org/10.1029/2019EF001382
22. Jha AK, Bloch R, Lamond J (2012) Cities and flooding. The World Bank
23. Juliá PB, Ferreira TM, Rodrigues H (2021) Post-earthquake fire risk assessment of historic urban areas: a scenario-based analysis applied to the Historic City Centre of Leiria, Portugal. Int J Disaster Risk Reduct 60:102287. https://doi.org/10.1016/j.ijdrr.2021.102287
24. Kappes MS, Keiler M, von Elverfeldt K, Glade T (2012) Challenges of analyzing multi-hazard risk: a review. Nat Hazards 64:1925–1958. https://doi.org/10.1007/s11069-012-0294-2
25. Lau CL, Smythe LD, Craig SB, Weinstein P (2010) Climate change, flooding, urbanisation and leptospirosis: fuelling the fire? Trans R Soc Trop Med Hyg 104:631–638. https://doi.org/10.1016/j.trstmh.2010.07.002

26. Leichenko R (2011) Climate change and urban resilience. Curr Opin Environ Sustain 3:164–168. https://doi.org/10.1016/j.cosust.2010.12.014
27. Lindell MK, Arlikatti S, Prater CS (2009) Why people do what they do to protect against earthquake risk: perceptions of hazard adjustment attributes. Risk Anal 29:1072–1088. https://doi.org/10.1111/j.1539-6924.2009.01243.x
28. Maio R, Estêvão JMC, Ferreira TM, Vicente R (2017) The seismic performance of stone masonry buildings in Faial island and the relevance of implementing effective seismic strengthening policies. Eng Struct 141:41–58. https://doi.org/10.1016/j.engstruct.2017.03.009
29. Meerow S, Newell JP, Stults M (2016) Defining urban resilience: a review. Landsc Urban Plan 147:38–49. https://doi.org/10.1016/j.landurbplan.2015.11.011
30. Michiels TLG (2015) Seismic retrofitting techniques for historic adobe buildings. Int J Archit Herit 9:1059–1068. https://doi.org/10.1080/15583058.2014.924604
31. Milne J (1912) A catalogue of destructive earthquakes, AD 7-1899. British Association for the Advancement of Science, Portsmouth, United Kingdom
32. Olshansky R (1996) Planning for hillside development, Planning A. American Planning Association, Washington DC, USA
33. Pelling M (2012) Hazards, risk and urbanisation. In: Wisner B, Gaillard JC, Kelman I (eds) Handbook of hazards and disaster risk reduction. Taylor & Francis, London, United Kingdom, p 912
34. Pierce JC, Budd WW, Lovrich NP (2011) Resilience and sustainability in US urban areas. Env Polit 20:566–584. https://doi.org/10.1080/09644016.2011.589580
35. Radeloff VC, Helmers DP, Kramer HA et al (2018) Rapid growth of the US wildland-urban interface raises wildfire risk. Proc Natl Acad Sci 115:3314–3319. https://doi.org/10.1073/pnas.1718850115
36. Scawthorn C (1986) Fire following earthquake. Fire Saf Sci 1:971–979. https://doi.org/10.3801/IAFSS.FSS.1-971
37. Schuster RL, Highland LM (2007) The third hans cloos lecture. Urban landslides: socioeconomic impacts and overview of mitigative strategies. Bull Eng Geol Environ 66:1–27. https://doi.org/10.1007/s10064-006-0080-z
38. Sevieri G, Galasso C, D'Ayala D et al (2020) A multi-hazard risk prioritisation framework for cultural heritage assets. Nat Hazards Earth Syst Sci 20:1391–1414. https://doi.org/10.5194/nhess-20-1391-2020
39. Sharma A, Mittal H, Gairola A (2018) Mitigation of wind load on tall buildings through aerodynamic modifications: review. J Build Eng 18:180–194. https://doi.org/10.1016/j.jobe.2018.03.005
40. Sheikh H, Van Engelen NC, Ruparathna R (2022) A review of base isolation systems with adaptive characteristics. Structures 38:1542–1555. https://doi.org/10.1016/j.istruc.2022.02.067
41. Solecki W, Leichenko R, O'Brien K (2011) Climate change adaptation strategies and disaster risk reduction in cities: connections, contentions, and synergies. Curr Opin Environ Sustain 3:135–141. https://doi.org/10.1016/j.cosust.2011.03.001
42. Tilloy A, Malamud BD, Winter H, Joly-Laugel A (2019) A review of quantification methodologies for multi-hazard interrelationships. Earth-Sci Rev 196:102881. https://doi.org/10.1016/j.earscirev.2019.102881
43. UNISDR (2017) Flood hazard and risk assessment. words into action guidel natl disaster risk assess hazard specific risk assess, pp 1–16
44. United Nations (2019) Global assessment report (GAR2019). Washington DC, USA
45. Zlateski A, Lucesoli M, Bernardini G, Ferreira TM (2020) Integrating human behaviour and building vulnerability for the assessment and mitigation of seismic risk in historic centres: proposal of a holistic human-centred simulation-based approach. Int J Disaster Risk Reduct 43:101392. https://doi.org/10.1016/j.ijdrr.2019.101392
46. Zscheischler J, Martius O, Westra S et al (2020) A typology of compound weather and climate events. Nat Rev Earth Environ 1:333–347. https://doi.org/10.1038/s43017-020-0060-z

Chapter 2
Methods, Techniques, and Tools for Evaluating and Managing Risks in Urban Areas

José Carlos Domingues and Maria Xofi

Abstract In recent years, a significant amount of research has focused on evaluating the risks associated with individual natural hazards in urban areas. However, the cumulative effects of multiple hazards have not yet received widespread attention in the research community or have been widely incorporated into urban management practices. Considering that a better understanding of the impacts of cascading and compound hazards, including future hazards arising from climate change, is a fundamental requirement to support more focused and participated policy actions by stakeholders, this chapter discusses concepts such as urban risk and disaster resilience, as a way to introduce some of the challenges that appear while conducting a multi-hazard assessment. Indicator-based methodologies are presented as an interesting approach to multi-hazard assessment, and the importance of considering different levels of detail according to the scale of assessment is discussed.

Keywords Disaster risk · Risk management · Urban areas · Climate change · Indicator-based methodologies

1 Understanding the Interconnected Risks of Multiple Hazards in Urban Areas

Urban areas are increasingly vulnerable to the impacts of multiple hazards due to their high population densities, complex infrastructure systems, and increasing exposure to a range of natural, technological, and human-induced threats. While a considerable

J. C. Domingues (✉) · M. Xofi
Institute for Sustainability and Innovation in Structural Engineering (ISISE), Department of Civil Engineering, University of Minho, Guimarães, Portugal
e-mail: jcarl.domingues@gmail.com

M. Xofi
e-mail: m.xofi@utwente.nl

M. Xofi
Faculty of Engineering Technology, University of Twente, Enschede, The Netherlands

T. M. Ferreira (ed.), *Multi-risk Interactions Towards Resilient and Sustainable Cities*, Advances in Sustainability Science and Technology,
https://doi.org/10.1007/978-981-99-0745-8_2

volume of research has focused on evaluating the risks associated with individual hazards, the cumulative effects of multiple hazards have received relatively little attention in the research community or in urban management practices. However, this is starting to change as the importance of understanding cascading and compound hazards, including those arising from climate change, is increasingly recognized as a fundamental requirement to support more informed and participatory policy-making by various stakeholders, including multilateral agencies and local governments. This interest is clearly on the international agenda. The United Nations, for example, has pointed out the assessment of "multi-hazard disaster risks and the development of disaster risk assessments and maps, including climate change scenarios" as one of the priorities for action within the Sendai Framework for Disaster Risk Reduction 2015–2030 (Priority 1: Understanding disaster risk). Similarly, the European Union has made climate change adaptation, risk prevention, and resilience measures a strategic priority within the Europe 2020 Strategic Framework.

This renewed interest in multi-hazard risk assessment stems from the conclusion, in the aftermath of recent events, that existing disaster risk reduction frameworks are still inadequate to address events characterized by the simultaneous occurrence of disasters induced by natural hazards. In modern societies, where cities are interconnected and subject to the indirect disruptions caused by hazards in other regions, it is crucial to understand the ways in which different disasters can trigger or exacerbate one another, which requires a more comprehensive and integrated approach to assessing and managing urban risk, taking into account the complex interdependencies and feedback loops between different hazards and their impacts. This means considering not only the direct impacts of individual hazards, but also the indirect effects of one hazard on another and the ways in which different hazards can interact to create new or exacerbated risks. It also means recognizing that hazards do not affect all communities and populations equally, and that a more inclusive and participatory approach is needed to address the diverse and often intersecting vulnerabilities faced by different groups in urban areas.

2 Integrating Disaster Resilience and Sustainability in Urban Risk Management

As mentioned earlier, urban areas are increasingly vulnerable to the impacts of multiple hazards due to factors such as their high population densities, complex infrastructure systems, and increasing exposure to various threats, including natural threats, technological accidents, and human-induced risks. The built environment plays a critical role in shaping these risks and the ability of cities to withstand and recover from disasters. The design, construction, and maintenance of buildings, infrastructure, and public spaces can either increase or decrease vulnerability to hazards, as well as influence the resilience of urban systems and the sustainability of urban development.

The built environment of towns, cities, and historical centers is comprised of different building typologies and infrastructure systems (roads, power, water), that vary in quality and level of maintenance and can be impacted by hazards to different degrees. The effective capacity of building resilience, especially in highly populated areas, depends on a thorough understanding of their exposure to different hazards, which span from urban fires to floods, earthquakes and landslides or even tsunamis in coastal areas. In this regard, it is important to consider that the risks resulting from these hazards are strictly interconnected, as different threats can interfere among them and result in a so-called domino effect, posing additional challenges for emergency and post-disaster systems.

Since the definition and approval of the global framework for disaster risk reduction in Sendai (Japan) in 2015, there has been a growing recognition of the value and close relationship between disaster resilience and sustainability, and the need for policies and tools that support risk-informed sustainable development [31]. This is particularly important in urban and coastal areas, which are facing increased exposure and vulnerability to hazards due to climate change [21]. In addition, and not less importantly, disasters have severe consequences for human societies, not only in terms of human lives and economic losses, but also in hindering human development. Figures might differ, but it is long agreed that while economic losses are often concentrated in more developed countries, most deaths are concentrated in less developed countries [5, 27]. To enhance the preparedness of cities to mitigate present and future risks arising from multiple hazards, it is therefore essential to develop integrated decision-support tools that can assess multiple hazards at the urban scale. Such tools can help prioritize resilience actions and preventive interventions and form the basis for exploring institutional adjustments that improve stakeholders' capacities to manage risk. They can also support post-disaster recovery efforts and prevent (or minimize) the adverse effects of hazardous events on people and the built environment, including the destruction of valued architectural and cultural assets.

3 On the Use of Qualitative and Continuous Approaches to Assess Urban Risk

Urban risk is a multi-dimensional matrix that includes multiple elements at risk (such as people, buildings, and infrastructures), multiple hazards (such as geophysical, meteorological, and hydrological), multiple temporal scales (including the present and future points in time), and multiple types of vulnerabilities. There are also direct risks, such as injuries and fatalities, and indirect consequences, such as disruptions to the functioning of urban areas, that pose a threat to their sustainable development, particularly to those located in the coastal areas. As a result, multi-hazard assessment must consider not only the different characteristics of individual hazards (such as their processes, metrics, and time-frames; see Kappes et al. [22, 23], Forzieri et al. [15]), but also their mutual interrelations (e.g., tsunami or landslide induced by earthquakes or floods triggered by extreme rainfall). Cascading hazards occur sequentially in time, with one hazard triggering or altering the processes and impacts of another subsequent hazard. Compound hazards, on the other hand, are two hazards occurring and impacting the same time and place, but with a triggering interrelation between them. Recent scientific advances provide valuable insights into the historical and theoretical interactions between different types of hazards and the territorial actions that either impede or catalyze them. The research of Tilloy et al. [30] and Gill and Malamud [17, 18], on multi-hazard interrelationships, for example, is crucial for understanding the processes and impacts of cascading and compound hazards.

One of the major challenges in multi-hazard risk assessment is how to compare multiple hazards with different reference units (such as nature, intensity, or return period) and complex interrelationships. This complexity might be why, despite its importance, multi-hazard assessment is still an emerging field of research. Additionally, the complexity of available models may also be a barrier to their wider adoption by practitioners, as overly complex models and user-unfriendly tools can be difficult to use [24].

Considering different approaches and tools that might be used in order to address the all too known technical challenges arising in multi-hazard assessment, two main types of approaches may be identified: the qualitative and the continuous (or semi-quantitative) approaches.

In qualitative approaches, intensity and frequency thresholds are defined to classify hazards into a predefined number of classes. While these methods can theoretically assure compatibility between different hazards (e.g., high earthquake and high flood hazard), it can be challenging to compare information from different sources, as different criteria may have been used to define the thresholds. Examples of qualitative approaches include the Swiss Guidelines for the Analysis and Evaluation of Natural Hazards [20], the French Risk Prevention Plans [8], and the ARMONIA project [9].

Continuous approaches, on the other hand, use indices (or a small number of indices) to represent risk. In contrast to qualitative approaches, indices offer a continuous standardization of different parameters (which may not be directly comparable, though) and allow for the quantification of the difference between two hazard levels rather than simply ranking them. Inductive approaches, a type of continuous approach, model risk by weighting and combining different hazard, vulnerability, and risk reduction factors, focusing on characteristics that make people and cities vulnerable to natural and human-made events. Inductive approaches are particularly useful in locations without empirical data on damage and the magnitude of processes [1]. In addition, their flexibility and adaptability make them suitable for climate change scenarios, as they can be easily modified to account for changing hazard and vulnerability profiles. Examples of continuous approaches include the Integrated Rapid Visual Screening (IRVS) and the Multi-hazard City Risk Index (MHCRI). Developed by the United States Department of Homeland Security, the IRVS [10] is a software-facilitated procedure intended to be used to identify the level of risk for a single building, to identify the relative risk among buildings in a community or region, and to set priorities for further risk management activities. As for the MHCRI, it was developed by the World Bank to measure multiple risks at the city level from natural hazards with both rapid and gradual onset [29]. Both these approaches demonstrate that a single index can be an effective, adequate, and useful basis for supporting policy decisions aimed at reducing urban risks and vulnerabilities and improving resilience.

Although index-based methodologies can effectively integrate the various dimensions of urban risk, the truth is that most previous research on multi-hazard risk assessment has analyzed individual hazards separately, i.e., without considering their interconnectedness and potential for complex interactions. As already addressed in this chapter, this assumption is not always valid, as natural processes are components of complex systems, and therefore they are not completely independent. Instead, they are interconnected, interacting in nonlinear ways to create new and diverse hazard patterns.

Moreover, most of these indices (both for individual and multiple hazards) have been designed to measure risk at the country scale, with only a few being developed for the urban scale. This is a significant limitation, as urban areas often face unique risk profiles and may require more tailored risk assessment approaches. Not less important, to date, there is currently a lack of approaches in the literature that integrate both disaster risk reduction (DRR) and climate change adaption (CCA) in the context of urban risk assessment. This is a particularly pressing issue, as, as mentioned earlier, climate change is expected to exacerbate many existing hazards and introduce new ones, making it increasingly important to consider the impacts of both current and future hazards.

4 Addressing Vulnerability as a Key Component of Urban Risk

As has been demonstrated by several past research works devoted to this topic [12], the range of options available to local governments to reduce risks is still limited. Given current scientific knowledge, it is not yet possible to alter the magnitude or frequency of natural hazards. Likewise, it is difficult to significantly reduce the number of elements at potential risk, either because these elements are often closely interconnected and interdependent, and efforts to reduce the risk to one element may have unintended consequences for other elements, or because they may be difficult or, in some cases, impossible to relocate or protect. However, governments can influence vulnerabilities by reducing the physical susceptibility of the exposed elements by improving the structural quality of buildings and infrastructures, controlling settlement in hazard-prone areas, and investing in public open spaces that can be safe post-disaster collection and distribution areas. Governments can also reduce fragility by taking steps to anticipate impacts, educate, and inform vulnerable populations about disaster risk management and climate change adaptation, invest in strategies that improve the capacity of these populations to respond to hazards, and enhance urban resilience. This can involve implementing early warning systems, conducting risk assessments, and developing contingency plans, providing information about how to prepare for and respond to hazards and raising awareness about the importance of resilience; providing training and necessary resources or equipment; and improving institutional coordination and strengthening human capital in disaster risk management.

Overall, addressing vulnerability involves considering three key variables:

- Susceptibility, which refers to physical conditions. In a simple way, this can be thought of as a measure of the amount of damage that a specific hazard causes, taking into account the physical characteristics of an asset;
- Fragility, which is based on social, economic, and financial parameters (such as social conditions, urban economy, governance, financial capacities, etc.). It is a proxy measure of the potential for damage that could be caused by the hazard. Fragility parameters are an attempt to identify relative capacities to prepare for and withstand hazardous events. Fragility parameters should be applicable to all hazards, rather than being specific to any one hazard; and
- Resilience, which evaluates the ability of the urban area to prevent, withstand, mitigate, and recover from a damaging event. Resilience is seen as a "positive variable", and therefore, unlike susceptibility and fragility, it has a negative weight in the final vulnerability measure.

In general, vulnerability is a relative measure of how much damage or harm may result to a community based on different factors, including physical susceptibility (of buildings and infrastructure), fragility (social, economic, and financial factors), and resilience (planning, regulation, Disaster Risk Management, and Climate Change Adaption measures). The overall analysis of vulnerability is a complex task, given that, while individual vulnerability elements can be subjected to a reasonably robust analysis, how different elements accumulate is a complex problem, exacerbated by their mutual dependencies. Besides, some vulnerabilities are considered to be generic, while others are very hazard-specific—for example, a building can be vulnerable to fire but not to an earthquake.

As already discussed in the context of urban risk assessment, index-based methodologies (which were previously used mainly in studies focusing on social vulnerability) can also be quite useful for quantifying individual physical vulnerability in the context of broader risk assessment frameworks. These methodologies have the advantage of being able to take into account a wide range of characteristics, which can be given different weights to reflect the distinct contributions of each feature to overall physical vulnerability. This is particularly useful, as some features may be more relevant in certain hazards or have different levels of importance in relation to different natural processes. Additionally, weights can be customized to fit the needs of the user; for example, a user focused on reducing pre-disaster vulnerability may assign different weights to characteristics than a user aiming to assess potential human and material losses post-disaster in order to prioritize emergency response.

From the exposed, one can conclude that index-based methodologies are more comprehensive than many other more widespread methodologies, such as vulnerability curves, which typically only consider a single characteristic of the building [28]. They are also flexible enough to be applied to disasters caused by various hazards, and used to address the combined impacts of multiple hazards. A further advantage of index-based methodologies is that they can be used in regions where there is no historical record of a specific type of disaster. This is especially relevant given the changes in hazard patterns caused by climate change and the prediction that previously uncommon natural hazards will affect regions worldwide. Index-based methodologies are also well-suited to integrating information from different sources, such as physical, social, and institutional factors.

For all these reasons, index-based methodologies can be seen as a powerful tool for spatial planning, vulnerability reduction, and post-disaster emergency management. They can be used, for example, to decide which buildings should be retrofitted or receive emergency and evacuation efforts, as well as to assess the impact of measures aimed at mitigating building vulnerability and urban risk [2, 3, 12]. However, despite their undeniable potential, the use of index-based methodologies for physical vulnerability assessment is still in its early stages [1]. There are also questions that need to be addressed regarding the selection and weighting of indicators. Overall, it can be said that the use of index-based methodologies in this field is still developing.

5 Balancing Detail and Scale in Disaster Risk Assessment

Effective disaster risk reduction strategies must be based on a thorough understanding of the magnitude and frequency of the natural hazards relevant to the area being studied, as well as the dynamics of the affected communities and the vulnerability of the exposed buildings and infrastructures. With that in mind, the very first and critical task is to select the approach to adopt—a process that is not always as easy as it might sound. The first aspect that is necessary to consider is the differences between detailed vulnerability assessment approaches, typically more suitable for individual buildings, and more practical and cost-effective methodologies for larger areas, for which the large amount of data involved in the analysis poses important methodological challenges. In these cases, it may be more suitable to use less sophisticated and onerous inspection and recording tools, such as vulnerability assessment methods based on a few empirical parameters. This section addresses some of the issues related to the scale of assessment. It provides a brief overview of different seismic vulnerability assessment methods, which will be used to illustrate the different levels of detail that must be considered when assessing the vulnerability of exposed elements at scales—ranging from an individual building to the national level, regardless of the natural hazard being considered.

Over the last few decades, there has been increasing interest in assessing the performance of buildings under natural hazards due to growing public awareness of the need to safeguard human life and protect architectural heritage. To minimize losses and reduce potential damage in the structures, several studies have been conducted at different scales to assess vulnerability and risk. The ultimate goal of these studies is to reduce potential damage and, this way, avoid or minimize losses against future hazardous events [4].

As mentioned above, the selection of the most appropriate seismic vulnerability assessment method should be based on a delicate balance between the simplicity of the methodology and the accuracy of the results, taking into account all the types of uncertainty that may potentially affect the analysis. Those include, for example, limitations of the model itself and inherent randomness in both the sample and response.

In the past decade, several European consortiums have focused on various aspects of vulnerability and seismic risk assessment and mitigation, including the classification and categorization of existing approaches at different scales. As discussed in detail by [13], these methodologies can be grouped based on two fundamental criteria: level of detail and quality of the input data, and type of output. According to the first criterion, vulnerability assessment approaches can be grounded into three different categories as a function of their level of detail:

- First-level approaches, which primarily use qualitative information and are, therefore, particularly suitable for large-scale analysis;
- Second-level approaches, which rely on mechanical models and detailed geometrical and mechanical information and are well-suited for the assessment of small building stocks (aggregates or single buildings); and
- Third-level approaches, which utilize complex numerical modeling techniques that require comprehensive and thorough geometrical, material, and mechanical information. Therefore, third-level approaches are only suitable for assessing individual, very well-characterized buildings.

One important component of vulnerability assessment approaches, considering the need to optimize human and technical resources in light of the scale of the assessment, lies in effectively obtaining and managing amounts of data that, at the urban scale, can become quite substantial. The use of carefully tailored datasheets, as a way to organize the gathering of the data obtained during on-site inspections, is a useful strategy towards the success of the methodology to use. As was pointed out by Ferreira and Ramirez Eudave [14], the design of the survey is a critical step in the adequacy and success of vulnerability assessment approaches. Ferreira and Ramirez Eudave [14] address as well different tools that can be used to complement and improve on-site inspections, from GIS platforms to remote sensing (see also Columbro et al. [6]).

The second criterion mentioned above refers to the type of output, or scale of the evaluation, which distinguishes the existing vulnerability assessment approaches into three main groups: direct, indirect, and hybrid techniques. The main difference between these three types of techniques lies in the number of steps involved in the analysis.

Direct techniques use a one-step process to estimate the structural damage caused by a certain hazard, and include two types of methodologies, known in the literature as typological and mechanical methodologies. Typological methodologies classify each structure within the building stock into typological classes considering a series of characteristics (such as structural, construction, material) that rule the ability of each building typology to cope with that hazard. A well-known example of this type of methodology is the Damage Probability Matrices (DPMs) approach, which analyzes data collected from post-event damage observations to determine the likelihood of damage for different magnitudes of hazard and building typologies [13, 26]. Mechanical methodologies, on the other hand, are used to predict the response of the structure by using analytical models of the building. Simplified mechanical models are typically more suitable for analyzing large sets of buildings with few input parameters while requiring modest computing time and effort and providing reliable quantitative results. An example of this type of methodology, applied to seismic vulnerability, is the FAMIVE method [7], which takes into account various mechanisms related to structural and constructional features of the building to characterize its vulnerability. More complex mechanical methodologies, such as the Capacity Spectrum [16] or the N2 method [11], are used to evaluate individual buildings at a higher level of detail, using more refined modeling techniques and analyses.

Indirect techniques, on the other hand, require a two-step process to estimate the damage and are used extensively in vulnerability assessment of urban areas and large building stocks. The Vulnerability Index Method is one of the most applied indirect methods and aims to characterize the physical and structural characteristics of the building in a quantitative form. To do this, a seismic vulnerability index is obtained in the first phase, either by using pre-available data (i.e., census data or municipality archives) or by carrying out on-site inspections. In the second phase, the damage associated with each building is estimated by using existing statistically based correlations derived from post-event observed damage data [26]. Several researchers have applied empirical vulnerability index-based methodologies using either the GNDT or European macroseismic approaches in many urban cities to obtain the seismic risk assessment. According to Ferreira et al. [13], the main advantage of this method is that it determines the vulnerability characteristics of the building stock rather than a certain building typology, which may or may not be representative of the actual building stock. However, provided that vulnerability indices are defined based on empirical-based coefficients (weights), there is a certain degree of uncertainty associated with the results and should therefore be taken into consideration and be able to validate or calibrate the seismic vulnerability results. Hence, indirect techniques (such as the vulnerability index-based method) should be applied to data gathered from a large number of buildings (at an urban or national scale) to be representative of the building stock. In addition to indirect methods, mechanical (or numerical) procedures are used to evaluate the seismic vulnerability of structures and can be distinguished between simplified methods with low computational effort and more complex methods, which require more detailed analysis and higher computational skills [26]. These methodologies require a large amount of information and are suitable to areas for which construction details are recorded, well understood, and often combined with some experimental work to characterize their mechanical behavior and recorded damage. In such a case, the information can be used then not only to calibrate the procedure itself but also to provide appropriate strengthening strategies at the level of the single building, urban block and district or entire city [32].

Finally, hybrid techniques combine both direct and indirect technique features, as described by Maio et al. [26]. An example of a hybrid technique, also applied to seismic vulnerability, is the macroseismic methodology proposed by Lagomarsino and Giovinazzi [25]. This approach combines the characteristics of typological and indirect techniques by using the vulnerability classes defined according to the EMS-98 scale [19] and a vulnerability index.

6 Final Remarks

As a field that combines information from different realities, multi-hazard assessment presents a number of challenges that are widely recognized in the literature. While still in its infancy, indicator-based methodologies based on the selection and weighting of indicators representing different features of the building stock arise as a flexible and powerful approach that can be an appropriate tool as a way to optimize resources when conducting risk assessment studies on a larger scale. Its flexibility is also an advantage considering climate change and the need to simulate different scenarios for the assessment of risk.

When dealing with a large number of buildings over a national or urban scale, the resources and quantity of information required are significant, and thus the use of less sophisticated techniques or tools is more practical and necessary. From this, methodologies for vulnerability assessments at a large scale should be based only on a few parameters, some of empirical nature. The multi-hazard assessment can finally be used by relevant bodies such as planners, infrastructure managers, and emergency responders considering different scenarios of impact and providing effective risk mitigation plans at a municipality or parish level.

Acknowledgements The project "MIT-RSC—Multi-risk Interactions Towards Resilient and Sustainable Cities" (MIT-EXPL/CS/0018/2019) leading to this work is co-financed by the ERDF—European Regional Development Fund through the Operational Program for Competitiveness and Internationalization—COMPETE 2020, the North Portugal Regional Operational Program—NORTE 2020 and by the Portuguese Foundation for Science and Technology—GCT under the MIT Portugal Program at the 2019 PT call for Exploratory Proposals in "Sustainable Cities".

References

1. Agliata R, Bortone A, Mollo L (2021) Indicator-based approach for the assessment of intrinsic physical vulnerability of the built environment to hydro-meteorological hazards: review of indicators and example of parameters selection for a sample area. Int J Disaster Risk Reduct 8:50–67. https://doi.org/10.1016/j.ijdrr.2013.12.006

2. Aguado JLP, Ferreira TM, Lourenço PB (2018) The use of a large-scale seismic vulnerability assessment approach for masonry façade walls as an effective tool for evaluating, managing and mitigating seismic risk in historical centers. Int J Archit Herit 12:1259–1275. https://doi.org/10.1080/15583058.2018.1503366
3. Anglade E, Giatreli AM, Blyth A et al (2020) Seismic damage scenarios for the Historic City Center of Leiria, Portugal: analysis of the impact of different seismic retrofitting strategies on emergency planning. Int J Disaster Risk Reduct 44:101432. https://doi.org/10.1016/j.ijdrr.2019.101432
4. Bernardo V, Sousa R, Candeias P, Costa A, Campos Costa A (2021) Historic appraisal review and geometric characterization of old masonry buildings in lisbon for seismic risk assessment. Int J Archit Heritage, 1–21. https://doi.org/10.1080/15583058.2021.1918287
5. Chaudhary M, Piracha A (2021) Natural disasters—origins, impacts, management. Encyclopedia 1:1101–1131. https://doi.org/10.3390/encyclopedia1040084
6. Columbro C, Ramirez Eudave R, Ferreira TM, Lourenço PB, Fabbrocino G (2022) On the use of web mapping platforms to support the seismic vulnerability assessment of old urban areas. Remote Sens 14:1424https://doi.org/10.3390/rs14061424
7. D'Ayala D, Speranza E (2003) Definition of collapse mechanisms and seismic vulnerability of historic masonry buildings. Earthq Spectra, 19. https://doi.org/10.1193/1.1599896
8. Delattre A, Garancher T, Rozencwajg C, Touret T (2002) Jurisque-prévention des risques naturels. Technical Report 3, Ministere de l'Écologie et du Développement durable, France
9. Delmonaco G, Margottini C, Spizzichino D (2006) ARMONIA methodology for multi-risk assessment and the harmonisation of different natural risk maps. Deliverable 3(1):1
10. DHS (2011) Integrated rapid visual screening of buildings. Department of Homeland Security
11. Fajfar P, Gaspersic P (1996) The N2 method for the seismic damage analysis of regular buildings. Earthq Eng Struct Dyn 25:23–67. https://doi.org/10.1002/(SICI)1096-9845(199601)25:1%3c31::AID-EQE534%3e3.0.CO;2-V
12. Ferreira TM, Maio R, Vicente R (2017) Analysis of the impact of large scale seismic retrofitting strategies through the application of a vulnerability-based approach on traditional masonry buildings. Earthq Eng Eng Vib 16:329–348. https://doi.org/10.1007/s11803-017-0385-x
13. Ferreira TM, Mendes N, Silva R (2019) Multiscale seismic vulnerability assessment and retrofit of existing masonry buildings. Buildings 9(4):91. https://doi.org/10.3390/buildings9040091
14. Ferreira TM, Ramirez Eudave R (2022) Assessing and managing risk in historic urban areas: current trends and future research directions. Front Earth Sci 10:847959. https://doi.org/10.3389/feart.2022.847959
15. Forzieri G, Feyen L, Russo S, Vousdoukas M, Alfieri L, Outten S, Migliavacca M, Bianchi A, Rojas R, Cid A (2016) Multi-hazard assessment in Europe under climate change. Clim Change 137:105–119. https://doi.org/10.1007/s10584-016-1661-x
16. Freeman SA (1978) Prediction of response of concrete buildings to severe earthquake motion. American Concrete Institute SP-55:589–605
17. Gill JC, Malamud BD (2014) Reviewing and visualizing the interactions of natural hazards. Rev Geophys 52:680–722. https://doi.org/10.1002/2013RG000445
18. Gill JC, Malamud BD (2017) Anthropogenic processes, natural hazards, and interactions in a multi-hazard framework. Earth Sci Rev 166:246–269. https://doi.org/10.1016/j.earscirev.2017.01.002
19. Grünthal G, European Seismological Commission (Eds) (1998) European macroseismic scale 1998: EMS-98. European seismological commission, subcommission on engineering seismology. Working Group Macroseismic scales
20. Heinimann HR, Hollenstein K (1998) Methoden zur Analyse und Bewertung von Naturgefahren: eine risikoorientierte Betrachtungsweise. Umwelt-Materialien Nr.85, Bundesamt für Umwelt, Wald und Landschaft, Bern, Switzerland
21. IPCC (2018) Global warming of 1.5 °C. An IPCC Special Report on the impacts of global warming of 1.5 °C above pre-industrial levels and related global greenhouse gas emission pathways, in the context of strengthening the global response to the threat of climate change, sustainable development, and efforts to eradicate poverty. https://www.ipcc.ch/sr15/

22. Kappes MS, Gruber K, Frigerio S, Bell R, Keiler M, Glade T (2012) The MultiRISK platform: the technical concept and application of a regional-scale multi-hazard exposure analysis tool. Geomorphology 151–152:139–155. https://doi.org/10.1016/j.geomorph.2012.01.024
23. Kappes MS, Keiler M, Elverfeldt K, von Glade T (2012) Challenges of analyzing multi-hazard risk: a review. Nat Hazards 64:1925–1958. https://doi.org/10.1007/s11069-012-0294-2
24. Komendantova N, Mrzyglocki R, Mignan A, Khazai B, Wenzel F, Patt A, Fleming K (2014) Multi-hazard and multi-risk decision-support tools as a part of participatory risk governance: feedback from civil protection stakeholders. Int J Disaster Risk Reduct 8:50–67. https://doi.org/10.1016/j.ijdrr.2013.12.006
25. Lagomarsino S, Giovinazzi S (2006) Macroseismic and mechanical models for the vulnerability and damage assessment of current buildings. Bull Earthq Eng 4(4):415–443. https://doi.org/10.1007/s10518-006-9024-z
26. Maio R, Ferreira TM, Vicente R (2018) A critical discussion on the earthquake risk mitigation of urban cultural heritage assets. Int J Disaster Risk Reduct 27:239–247. https://doi.org/10.1016/j.ijdrr.2017.10.010
27. O'Brien G, O'Keefe P, Rose J, Wisner B (2006) Climate change and disaster management. Disasters 30:64–80. https://doi.org/10.1111/j.1467-9523.2006.00307.x
28. Papathoma-Köhle M (2016) Vulnerability curves vs. vulnerability indicators: application of an indicator-based methodology for debris-flow hazards. Nat Hazards Earth Syst Sci 16:1771–1790. https://doi.org/10.5194/nhess-16-1771-2016
29. The World Bank (2011) Methodology report: calculating multi-hazard city risk-resilient cities: multi-hazard city risk Index. The World Bank, Washington DC
30. Tilloy A, Malamud BD, Winter H, Joly-Laugel A (2019) A review of quantification methodologies for multi-hazard interrelationships. Earth Sci Rev 196:102881. https://doi.org/10.1016/j.earscirev.2019.102881
31. UNISDR (2015) Sendai framework for disaster risk reduction 2015–2030. United Nations Office for Disaster Risk Reduction
32. Vicente R, D'Ayala D, Ferreira TM, Varum H, Costa A, Silva JARM, Lagomarsino S (2014) Seismic vulnerability and risk assessment of historic masonry buildings. In: Costa A, Guedes JM, Varum H (eds) Structural rehabilitation of old buildings, vol 2. Springer, Berlin, Heidelberg, pp 307–348. https://doi.org/10.1007/978-3-642-39686-1_11

Chapter 3
Social Vulnerability in the Lisbon Metropolitan Area

Pedro Pinto Santos and Tiago Miguel Ferreira

Abstract The manifestation of a hazardous process in a given location is clear evidence of a threat to individuals and communities. Without hazard, there is no risk. Vulnerability, however, plays a less evident role in explaining the losses that are observed in databases, whether global or local. Social vulnerability, in particular, represents the underneath conditions that turn individuals and communities more or less able to endure the impacts of hazardous events. A detailed-level analysis of social vulnerability was performed in the Lisbon Metropolitan Area, considering the dimension of the individuals' characteristics—that we define as criticality—and the characteristics of the surrounding territories in the ability to provide support during and timely recovery after the event—that we define as support capability. The study area is highly contrasting in terms of this later dimension, with urban areas concentrating most of the services and equipment that reduce vulnerability. Regarding criticality, the methodology allowed to identify very-localized hotspots laid out to high propensity to losses from two drivers: employment and education (first principal component of criticality) and age, gender, and old urban fabric (second principal component). Analysed separately or combined in a single social vulnerability index, this information is useful in the planning of short-term actions in the strict field of civil protection operations and in mid- to long-term actions considering a wider perspective of risk governance, bringing to the table public policies in the areas of social care, mobility, urban planning, education, and health services, that address the very deep roots of vulnerability.

Keywords Social vulnerability · Criticality · Support capability · Risk assessment · Risk management

---

P. P. Santos (✉)
Centre for Geographical Studies, Institute of Geography and Spatial Planning (IGOT), LA TERRA, University of Lisbon, Lisbon, Portugal
e-mail: pmpsantos@campus.ul.pt

T. M. Ferreira
School of Engineering, College of Arts, Technology and Environment (CATE), University of the West of England (UWE Bristol), Bristol, UK
e-mail: Tiago.Ferreira@uwe.ac.uk

T. M. Ferreira (ed.), *Multi-risk Interactions Towards Resilient and Sustainable Cities*, Advances in Sustainability Science and Technology,
https://doi.org/10.1007/978-981-99-0745-8_3

1 Introduction

Vulnerability assessments are a keystone in the analysis of adaptation strategies not only for climate change [6, 16] but for disaster risk reduction (DRR) strategies as well, which is recognized as one of the four accelerators for the Sendai Framework implementation [17]. In fact, both global determinations—climate change adaptation and DRR—feature immense synergies and can only be effective if aligned at all scales and levels of intervention.

Social vulnerability is defined as the propensity of individuals, communities, and systems to be negatively affected by hazardous processes of diverse nature, based on their social and demographic characteristics [1, 2, 8, 10, 11]. Levels of social vulnerability act as predictors of the capacity of individuals, communities, and systems to cope with and recover from *inter alia*, the impact of disasters induced by natural processes [4], and are equally applicable to natural-induced or pandemic crisis situations such as the one experienced (to be still experienced?) recently. In this sense, social vulnerability and resilience are related as concepts and usefulness in risk assessment and management processes [3].

Social vulnerability (SV) is a key indicator for risk governance, involving the processes and impacts resulting from events of natural, technological, and environmental origin. The relevance of the analysis of social vulnerability at a sub-municipal level arises from the instrumental need to base policy options translated into strategic and operational measures in the context of risk prevention, reduction, mitigation, and adaptation [7], as well as the need to consolidate the indicators of support capability and community resilience. Therefore, this indicator responds to requirements in areas such as civil protection and emergency planning, social, health, and education policies, as well as contributing to the urban and spatial planning reference frameworks.

Current challenges in social vulnerability assessment lay on the ability to produce timely comparable vulnerability scores [5, 15], to find validation methodologies and data [13], to tailor the assessments to particular types of hazards [12], for instance in regard to flooding), and to incorporate socioeconomic projections in the models [18].

The aim of this study is to assess social vulnerability based on the perspective of Mendes [9], i.e., incorporating the individual dimension (criticality) and the collective dimension (support capability) of the proneness to suffer loss and the ability to recover in a timely manner. This SV assessment—and of its main drivers as expressed and mapped by the principal components—was done for the Lisbon Metropolitan Area (LMA) at the fine scale of the statistical block. Specific objectives are the collection of base information in support of the housing and residents' characteristics; the quantification and description of the main drivers of SV; the identification of the main socially vulnerable areas; and the provision of information to support the definition of local and regional policies for SV reduction—in the domains of health, elderly population, education, civil protection, urban planning, mobility, environment, and social assistance), acting upon the particular and most relevant SV drivers in any particular neighbourhood, city centre, or village in the LMA.

2 Study Area and Units of Analysis in the SV Assessment

The Lisbon Metropolitan Area (LMA) comprises 18 municipalities and 118 civil parishes. The resident population was, at the time of the SV assessment, 2,813,000 inhabitants, living in around 3,105 km^2 (a population density of 906 inhabitants./km^2). However, vast portions of the LMA are under ecological legal protection (marshland and nidification areas near the Tagus estuary), as well as the mountainous areas of Sintra (covering parts of the Sintra, Cascais, and Mafra municipalities) and Arrábida (covering parts of the Setúbal and Sesimbra municipalities).

In the LMA, there are 4521 statistical blocks (Fig. 1). They differ significantly in area and population, with urban areas being represented by smaller and more densely inhabited blocks, and rural and natural protected areas with vaster areas and a small number of inhabitants. There are two statistical blocks with zero inhabitants: in these cases, criticality is assigned the lowest score, but support capability is calculated likewise the units of analysis. On average terms, the mean area of statistical blocks is 0.64 km^2 and the mean population is 624 persons, a figure that, considering the LMA population of more than 2.8 million, expresses the high level of detail of the SV assessment.

3 Methodology for the Social Vulnerability Assessment

Social vulnerability (*SV*) was evaluated for the entire LMA using the census statistical block level, and it is the result of the product of criticality (*Cr*) and support capability (*SC*) [10], as formulated in Eq. **1**.

$$SV = Cr \times (1 - SC) \tag{1}$$

As intermediate results, for each of the 4521 statistical blocks that compose the LMA, scores are calculated for each principal component (PC) within criticality and support capability, in addition to the final SV score itself. The SV index aims to overcome the constraint of subnational scales of analysis by incorporating a territorial perspective in support capability dimension of SV. The proposed method incorporates not only the individual characteristics of the population and risk groups as it also considers the territorial context in which they are supported, i.e., the public and private equipment, infrastructure, and services that might play a role in attenuating losses and enhancing recovery [9].

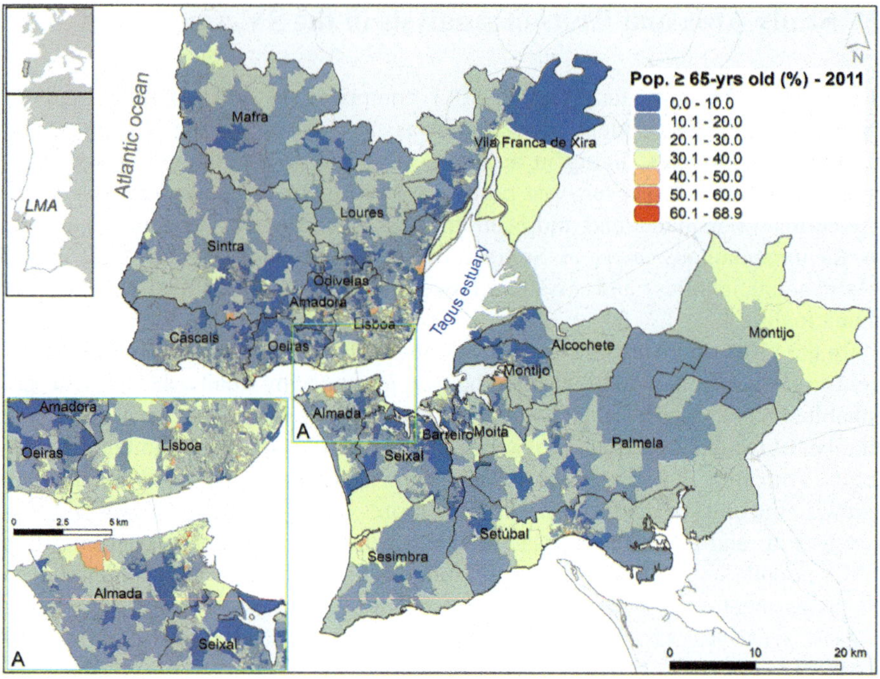

Fig. 1 Illustration of the fine scale of representation allowed by the selection of statistical block as the unit of analysis in the social vulnerability assessment. Representation of the proportion of population with or over 65 years old in the Lisbon Metropolitan Area

3.1 Criticality

Criticality (*Cr*) expresses the characteristics of individuals that make them prone to loss, considering their age, socioeconomic condition, health and housing conditions, social assistance, mobility, educational level, and employment. Data regarding an initial set of 43 variables was collected to perform the assessment of criticality (Table 1), which attempts to cover directly or indirectly those domains of vulnerability. In order to be comparable among statistical blocks, values of variables are expressed as proportions of the absolute value of the base variable (this means a % of the total residents, total dwellings, total buildings, etc.).

3.2 Support Capability

Support capability (*SC*) expresses the set of systems, networks, public and private infrastructures, and collective equipment aimed at supporting communities and their activities, which, in the eminence or occurrence of a dangerous process, make it

Table 1 List of variables used in the criticality assessment in the LMA

Code	Description of the variable
Vacant	Vacant dwelling units (%)
No Water	Dwelling units without water supply (%)
No Sewage	Dwelling units without sewage system (%)
Rented	Rented dwelling units (%)
1 or 2 Div	Dwelling units with 1 or 2 divisions (%)
5 plus Div	Dwelling units with five or more divisions (%)
Less 50 m^2	Dwelling units with less than 50 m^2 (%)
Plus 200 m^2	Dwelling units with more than 200 m^2 (%)
Plus 100 m^2	Dwelling units with more than 100 m^2 (%)
With Bath	Dwelling units with bathing facilities (%)
Owner	Dwelling units occupied by the owner (%)
Before 1970	Buildings built before 1970 (%)
1 or 2 floors	Buildings with 1 or 2 storeys (%)
5 plus floors	Buildings with five or more storeys (%)
Concrete	Buildings with a concrete structure (%)
Stone	Buildings with a structure of adobe and loose stone (%)
Study 9th	Individuals studying (1st 9th degree) (%)
Study muni	Individuals studying in the municipality of residence (%)
Complete 9th	Individuals with nine years of education completed (%)
Higher edu	Individuals with higher education completed (%)
Illiter	Illiterate individuals (%)
Primary sector	Individuals employed in the primary sector (%)
Secondary sector	Individuals employed in the secondary sector (%)
Tertiary sector	Individuals employed in the tertiary sector (%)
Unemployed	Individuals between 25 and 64 years old unemployed or looking for their first job (%)
Employed	Individuals between 25 and 64 years old employed (%)
Work muni	Individuals working in the municipality of residence (%)
Work study muni	Individuals working or studying in the municipality of residence (%)
No activ	Individuals without economic activity (%)
Work study out	Individuals working or studying outside the municipality of residence (%)
Indiv family	No. of individuals per family (no.)
5 plus family	Families with five or more elements (%)
1 or 2 family	Families with 1 or 2 elements (%)
65 plus family	Families with elements with 65 or more years old (%)
All employed	Families without unemployed elements (%)

(continued)

Table 1 (continued)

Code	Description of the variable
Child in family	Families with children less than 15 years old (%)
Pop 0 to 4	Individuals with less than five years old (%)
Pop 15 to 24	Individuals between 15 and 24 years old (%)
Pop 65 plus	Individuals 65 years old or older (%)
Indiv per dwell	No. of individuals *per* dwelling (no.)
Masculi rate	Masculinity rate (%)
Retired	Retired individuals (%)
Women pop	Women population (%)

possible to reinforce the community's capacity to mitigate and/or recover from a hazardous event. Common dimensions covered are the economic dynamism, the coverage by social equipment (for example, health centres), civil protection resources, and public and private businesses that provide essential goods and mobility. The equipment and services included in the assessment (Table 2) are expressed according to different methods of accounting for

- the coverage by fire stations, pharmacies, hospitals, health infrastructures, gas and power stations, and police stations is expressed by the distance from the centroid of the statistical block to the nearest entity;
- the coverage by the touristic equipment with capacity for temporary shelter expresses the sum of the lodging units in the hotels located inside or within 3 km from the boundaries of each statistical block;
- the coverage by the main road network is evaluated considering the location of the nearest nodes of the secondary and tertiary road hierarchy (node levels 3 and 4). This excludes node level 5 of connectivity (primary hierarchy, a level between highways only) and the urban node levels (1 and 2), which are less relevant in describing municipal and inter-municipal accessibility. The underneath rationale

Table 2 List of variables used in the support capability assessment in the LMA

Code	Description
Hotel housing	Coverage by hotels with capacity for temporary shelter
Fire sta	Coverage by fire stations
Pharm	Coverage by pharmacies
Road nodes 34	Coverage by the main road network
Hospital	Coverage by hospitals
Health centre	Coverage by health centres
Gas sta	Coverage by car gas and power stations
Police	Coverage by police stations
Grocery	Coverage by grocery stores

is to express i) the ability to be accessed by outside support (rescue and emergency operations, provision of essential goods) and ii) to be able to evacuate in emergency situations or to move in daily activities;

- the coverage by grocery stores also follows the nearest distance, but it also classifies the stores into two levels: class 1 for small bakeries, butcheries, and convenience stores; level 2 for municipal markets, supermarkets, and shopping centres.

All geographical entities are expressed as points, and they were collected using a buffer of 30 km from the boundaries of the LMA (except the grocery stores that followed a buffer of 7 km), in order to avoid the prejudice of near-boundary areas well covered by LMA-outside services and equipment. Unlike the data supporting the criticality assessment, which is based solely on Census information, the collection and integration of the geographical input data for the support capability assessment are far more time-consuming, and not entirely exempt from representation bias caused by the metrics in which the coverage is expressed.

The extent of the damage—for example, the number of casualties or the number of days with restricted mobility—will depend on the support capability of the territory. A high support capability may thus constitute a counterpoint to a high level of criticality. The location and density of infrastructures are a reflection of the way society is structured. While for a population with a high support capability, a certain damaging event may only take on fortuitous characteristics—since it has sufficient capacities and resources to be able to more or less easily restore the losses and damages suffered—in the case of a population framed in a territory with reduced support capability, that same event may mean the aggravation of existing fragilities, giving rise to situations of serious disruption of daily socioeconomic functions.

3.3 Statistical Procedure

Prior to the application of Eq. 1, autonomously for each dimension *Cr* and *SC*, the following steps are taken:

- normalization of data to the z-score;
- test of multicollinearity between input variables until a set of robust variables is achieved, excluding pairwise variables with Pearson correlation coefficients higher than 0.7;
- iterative application of Principal Component Analysis, using anti-image matrices' correlations, communalities' scores, KMO, and interpretation of the rotated component matrix to exclude unsuitable or irrelevant variables from the analysis, as well as additional redundant variables;
- after the final model is achieved, interpretation of the principal components and attribution of their cardinality, according to their role in explaining Cr and SC;

- sum of component scores with weighting defined by the % of variance explained by each principal component;
- linear transformation of values to an interval between 0 and 1;
- classification of the score of each unit of analysis according to its standard deviation.

The number of principal components (also named PCs) is defined from the eigenvalues above 1.

4 Results

4.1 Criticality

Following the statistical procedure described, the final PCA model for criticality was run with 12 variables, as presented in Table 3.

For that dataset, a KMO score of 0.722 is achieved, 73.7% of the total variance is explained, and four principal components (PC) with Eigenvalue >1 were extracted.

PCs represent the four drivers of criticality, as interpreted from the rotated components matrix (Table 4), and they were named as follows:

- employment and qualifications (PC1), which explains 32.5% of the total variance;
- age, gender, and ageing urban context (PC2), 22.5% of the total variance;
- housing conditions (PC3), 10.3% of the total variance;
- and family structure (PC4), 8.4% of the total variance.

Table 3 Final set of variables used in the criticality assessment, after redundancy elimination and analysis of robustness

Code	Variable	Communalities
(Rented)	Rented households (%)	0.802
(Less 50 m^2)	Dwelling units under 50 m^2 (%)	0.690
(With bath)	Dwelling units with bathing facilities (%)	0.590
(Before 1970)	Buildings built before 1970 (%)	0.725
(Concrete)	Building with a concrete structure (%)	0.617
(Complete 9th)	Individuals with nine years of education completed (%)	0.885
(Higher edu)	Individuals with higher education completed (%)	0.824
(Secondary sector)	Individuals employed in the (%)	0.751
(Employed)	Individuals between 25 and 64 years old employed (%)	0.692
(5 plus family)	Families with five or more elements (%)	0.812
(Pop 65 plus)	Individuals 65 years old or older (%)	0.815
(Women pop)	Women population (%)	0.644

Table 4 Rotated component matrix for the assessment of criticality. For the sake of interpretation, some variable names were simplified

	Principal components (PCs)			
	1	2	3	4
Pop. With Higher Education (%)	−0.889	0.119	−0.09	−0.111
Pop. With Elementary School (%)	0.860	0.289	0.246	0.031
Pop. Employed in the Industry (%)	0.717	−0.472	0.023	−0.117
Pop. Employed (%)	−0.703	−0.245	−0.096	−0.357
Pop. Over 65 years old (%)	0.233	0.822	0.188	−0.22
Women population (%)	−0.145	0.777	−0.122	−0.065
Buildings built before 1970 (%)	0.014	0.643	0.537	−0.152
Rented households (%)	0.154	0.549	0.511	0.464
Concrete buildings (%)	0.042	−0.091	−0.767	0.135
Households with bathroom (%)	−0.231	0.175	−0.708	−0.069
Households under 50 m^2 (%)	0.290	0.342	0.663	0.22
Families with five or more elements (%)	0.143	−0.311	−0.046	0.832
Cardinality	+	+	+	+
% of the total variance explained	32.492	22.474	10.315	8.446

The final cartographic expression of criticality at the statistical block in the LMA is represented in Fig. 2 and summarized in Figs. 3 and 4, and in Table 5.

The resident population in each statistical block can be summed according to the respective class of criticality in the entire LMA (Fig. 3) and by municipality (Table 5). An obvious remark to this summing is that not all residents with a unit of analysis feature the same levels of criticality. However, these figures represent a fair indication of the dominant levels of criticality.

A total of 150,649 inhabitants (5.3% of the population of 2,821,876) reside the in the 273 statistical blocks (6.0% of the total, 4521) classified with very high criticality. They are located particularly in some old neighbourhoods of Lisbon, as well as in some suburban areas in the municipalities of Almada, Amadora, Barreiro, Loures, Moita, and Setúbal. Very high criticality is rarely found in rural areas.

On the other side of the scale, very low criticality is assigned to 324 statistical blocks (7.2% of the total), where 217,214 inhabitants reside (7.7% of the total). In a simple generalization, these blocks are located in three types of geographical typologies: historically areas of low urban density located outside or in the near outskirts of the main cities (the examples of statistical blocks located in the Cascais, Sintra, and Oeiras municipalities); newly constructed areas of low density, frequently closed condominium (the case of some blocks in Cascais, Oeiras, Mafra, Montijo, Alcochete, and Almada); newly constructed areas of high urban density, i.e., tall buildings (exemplified by some neighbourhoods in northern Lisbon, eastern Lisbon, and Oeiras).

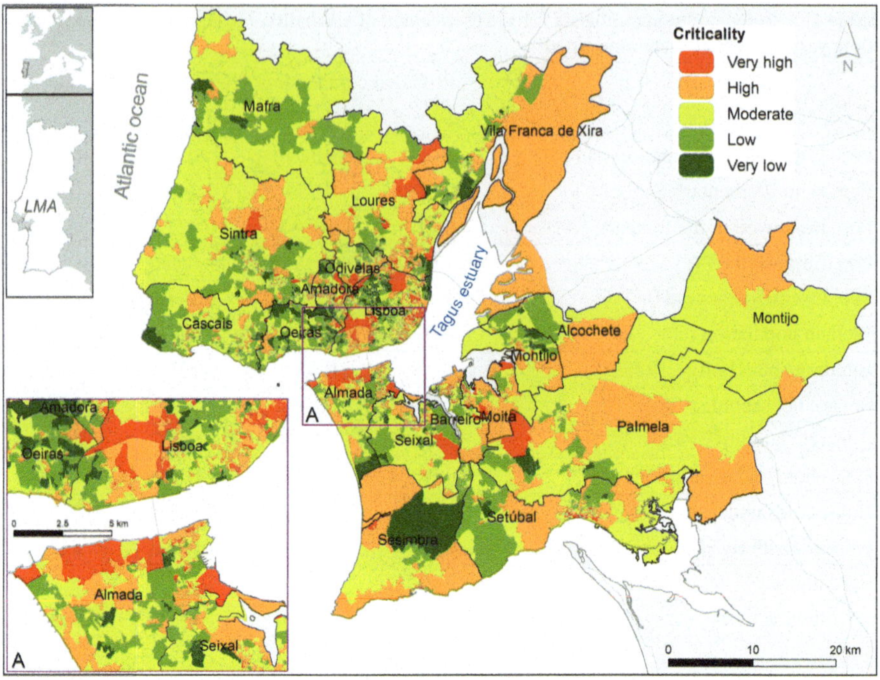

Fig. 2 Criticality in the Lisbon Metropolitan Area

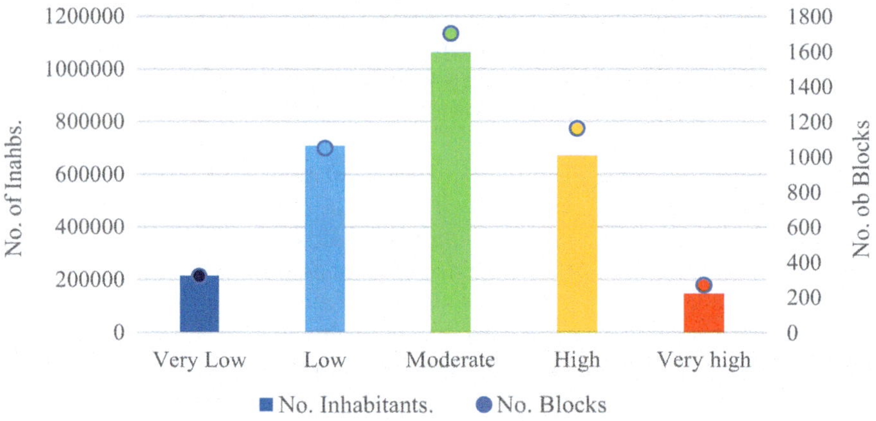

Fig. 3 Number of inhabitants and statistical blocks by class of criticality in the LMA

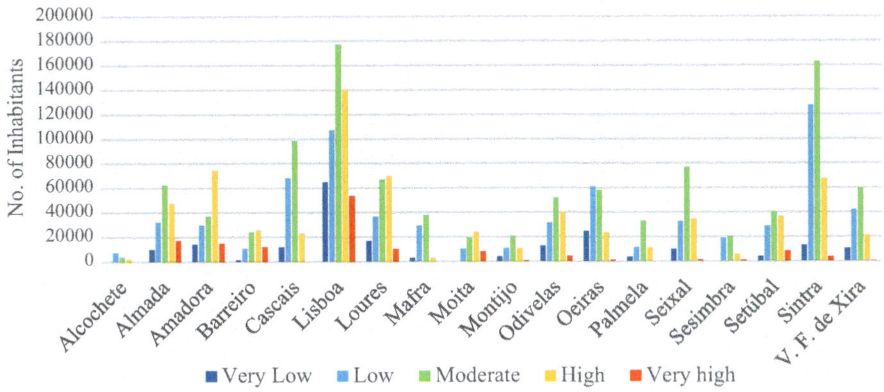

Fig. 4 Number of inhabitants by class of criticality in the LMA municipalities

Given the great detail of the analysis, considering the total of 4521 statistical blocks in the study area, the municipality of Sintra was chosen as an example to map the behaviour of each territorial unit in each of the criticality drivers (PC1–PC4); see Figs. 5, 6, 7 and 8.

Together, the four maps presented above show that a given territorial unit—whether a neighbourhood, a village, or a very small rural settlement between cities and villages—can feature high scores in one criticality driver and low scores in another. The combinations are multiple and define the criticality profile summarized in Fig. 2. For instance, the northern near-half portion of the municipality is essentially rural or newly urbanized areas where unemployment and low qualifications predominate (Fig. 5), which are not necessarily areas with bad housing conditions (poor housing conditions exist in the NE sector but not in the NW sector) (Fig. 7).

4.2 Support Capability

Following the statistical procedure described, the final PCA model for support capability was run with nine variables, as presented in Table 6.

For that set, an excellent KMO score of 0.912 is achieved, and 65.9% of the total variance is explained by the two principal components (PCs) with Eigenvalue >1 that were extracted. PCs represent the two drivers of support capability, as interpreted from the rotated components matrix (Table 7), and they were named as follows:

- general equipment and services coverage (PC1), which explains 54.7% of the total variance;
- coverage by equipment with capacity for temporary shelter (PC2), 11.1% of the total variance.

Table 5 Summary of the number of inhabitants and statistical blocks by class of criticality in the 18 municipalities of the LMA

Municipalities	Very low		Low		Moderate		High		Very high		Totals	
	No. blocks	No. inhab.	No. blocks	No. inhab.	No. blocks	No. inhab.	No. blocks	No. inhab.	No. blocks	No. inhab.	Blocks	Inhab.
Alcochete	1	477	12	8476	7	4897	6	3153	1	566	27	17,569
Almada	18	10,971	63	32,958	115	63,392	90	48,487	32	18,222	318	174,030
Amadora	26	15,172	46	30,578	58	38,083	121	75,450	28	15,853	279	175,136
Barreiro	3	2437	18	11,585	41	25,119	47	26,701	25	12,922	134	78,764
Cascais	17	12,626	106	69,022	156	99,957	40	24,174	1	700	320	206,479
Lisboa	113	65,526	213	108,169	361	178,282	274	141,469	93	54,287	1054	547,733
Loures	24	17,756	53	37,539	94	67,986	104	70,460	20	11,313	295	205,054
Mafra	5	3716	38	30,189	59	38,654	9	3796	1	330	112	76,685
Moita	0	0	13	11,307	30	20,674	40	24,980	18	9068	101	66,029
Montijo	6	4894	15	11,700	31	21,670	20	11,544	3	1414	75	51,222
Odivelas	18	13,383	38	32,308	68	52,781	66	41,005	8	5072	198	144,549
Oeiras	36	25,538	96	61,637	98	58,696	41	24,197	2	2052	273	172,120
Palmela	8	4233	16	12,074	50	33,898	20	11,996	2	630	96	62,831
Seixal	15	10,538	49	33,330	116	77,464	57	34,915	4	2022	241	158,269
Sesimbra	1	338	27	19,722	36	21,197	17	6498	6	1745	87	49,500
Setúbal	5	4528	41	29,303	63	40,923	63	37,264	20	9167	192	121,185
Sintra	15	13,883	158	127,997	237	163,791	111	67,849	7	4315	528	377,835
VF Xira	13	11,198	51	42,644	86	60,441	39	21,632	2	971	191	136,886
Total	324	217,214	1053	710,538	1706	1,067,905	1165	675,570	273	150,649	4521	2,821,876

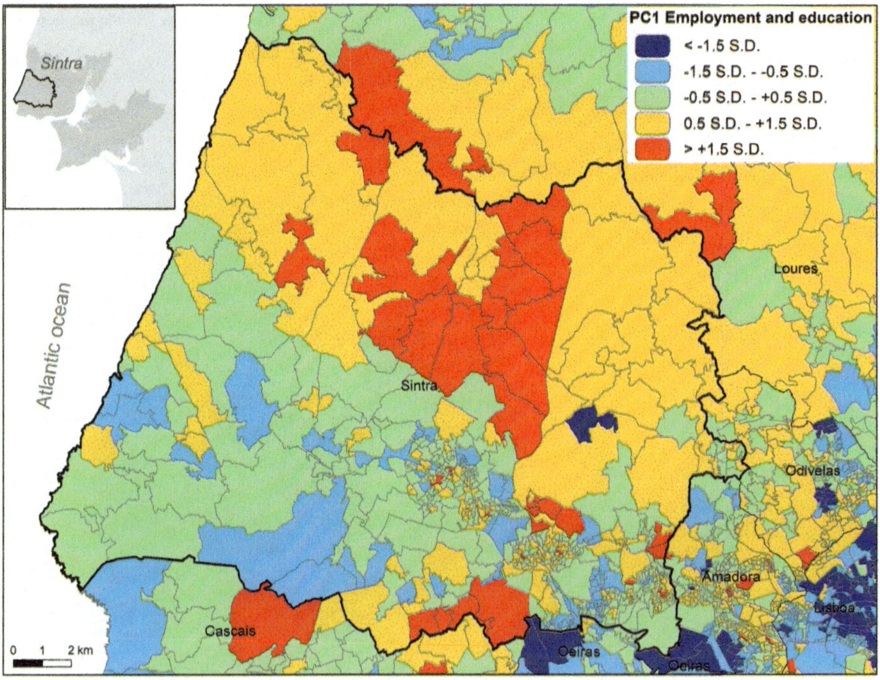

Fig. 5 Cartographic expression of the PC scores in regard to employment and education (PC1 of criticality) in the Sintra municipality

Cardinality in both PCs needed to be inverted because the shortest the distance to the equipment or service, the higher the support capability (the case of PC1, in which the explicative variables present a positive sign in the loading), and the higher the number of shelter units the higher the support capability (the case of PC2, in which the strongest explicative variable presents a negative sign in the loading).

The final cartographic expression of support capability at the statistical block in the LMA is represented in Fig. 9.

The final cartographic expression of support capability at the statistical block in the LMA is represented in Fig. 9 and summarized in Figs. 10 and 11, and Table 8.

The resident population in each statistical block can be summed according to the respective class of support capability in the entire LMA (Fig. 10) and by municipality (Table 8). As expressed in regard to criticality, naturally not all residents within a given class are characterized by it, despite the dominance of the class in the block.

A total of 726,600 inhabitants (25.7% of the population) reside in the 1308 statistical blocks (28.9% of the total) classified with very high support capability. This dimension is SV features a macrocephalic pattern spreading from the capital city of Lisbon to the metropolitan area. In the northern sector of LMA, starting from Lisbon, the spreading of areas with very high and high SC follows three alignments defined

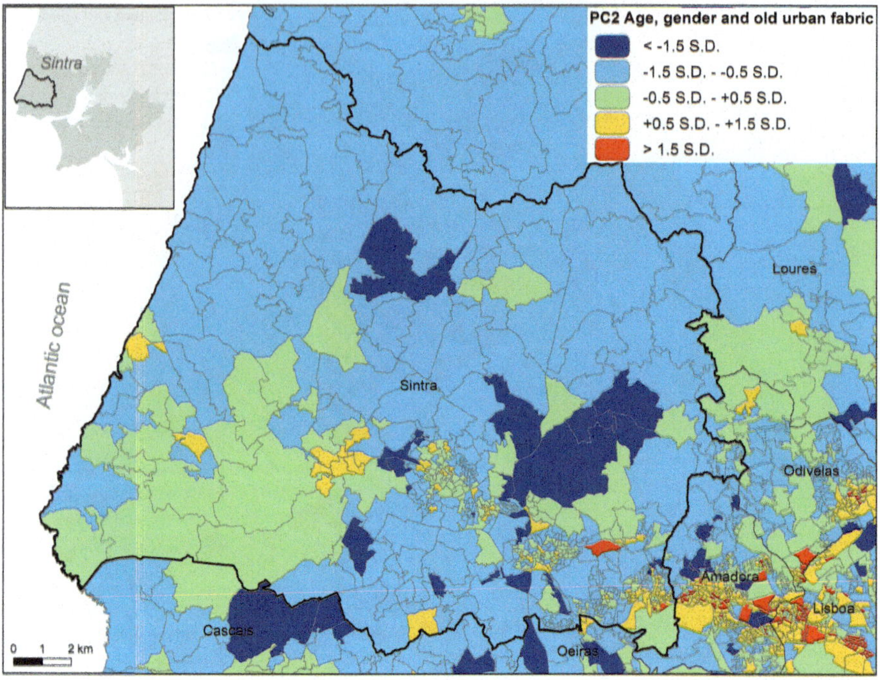

Fig. 6 Cartographic expression of the PC scores in regard to age, gender, and old urban fabric (PC2 of criticality) in the Sintra municipality

by the main communication routes (railway in the first phase, and road later): from Lisbon to V.F. Xira; from Lisbon to Sintra, including Amadora; and from Lisbon to Cascais, including Oeiras. Those alignments express the expansion of urbanization in the north LMA. In the south LMA, the pattern follows a similar process but in regard to the fluvial transportation and the location of the Tagus river bridge (between Almada and Lisbon). The city of Setúbal, by its proper economic dynamism, concentrates a high diversity and quantity of services and public equipment that result in high and very high support capability.

Very low scores of SC are found in the remaining interstitial areas, with small urban settlements, villages, or dispersed edification. They sum a total of 149 statistical blocks (3.3%) where 90,845 persons reside (3.2%).

The support capability drivers, expressed by the two principal components (PC1 and PC2), are mapped in Figs. 12 and 13.

As can be observed in the figures above, following the classification according to the standard deviation (S.D.), no statistical blocks in the LMA scored higher than 1.5 S.D. in regard to the general equipment and services coverage (PC1). Most blocks fall in the intermediate class (−0.5 to 0.5 S.D.), and it is concluded that the most urbanized areas, like Lisbon's city centre, are not necessarily the most covered by the

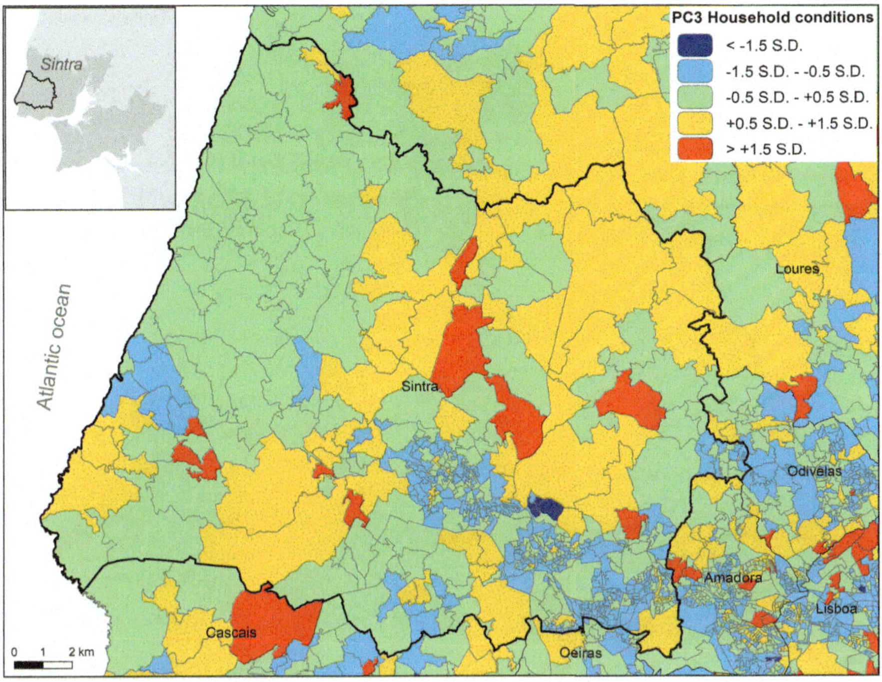

Fig. 7 Cartographic expression of the PC scores in regard to household conditions (PC3 of criticality) in the Sintra municipality

equipment and services included in the analysis. In terms of the temporary shelter provided by hotel facilities, the Lisbon municipality concentrates most of the lodging units available. According to the official data provided by the WebGIS SIGTUR of Tourism Portugal, the LMA presents a total of 63,962 lodging units—excluding camping areas and local lodgement provided in apartments—from which 26,902 (42.1%) are located in the Lisbon municipality. Less covered areas are located in the municipalities of Mafra, Sintra, Loures, Odivelas, Seixal, Sesimbra, and Palmela.

4.3 Social Vulnerability

As an expression of the product of criticality and support capability—as expressed in Eq. 1, and after the transformation to the amplitude [0, 1] of both dimensions—social vulnerability scores range between 0.00 and 0.55, with an average of 0.04. The three municipalities with the highest average are Palmela (0.11), Mafra (0.11), and Alcochete (0.09). On the opposite side, they are Lisbon, Amadora, and Oeiras (both with an average score of 0.02). Apart from the municipal average values, it

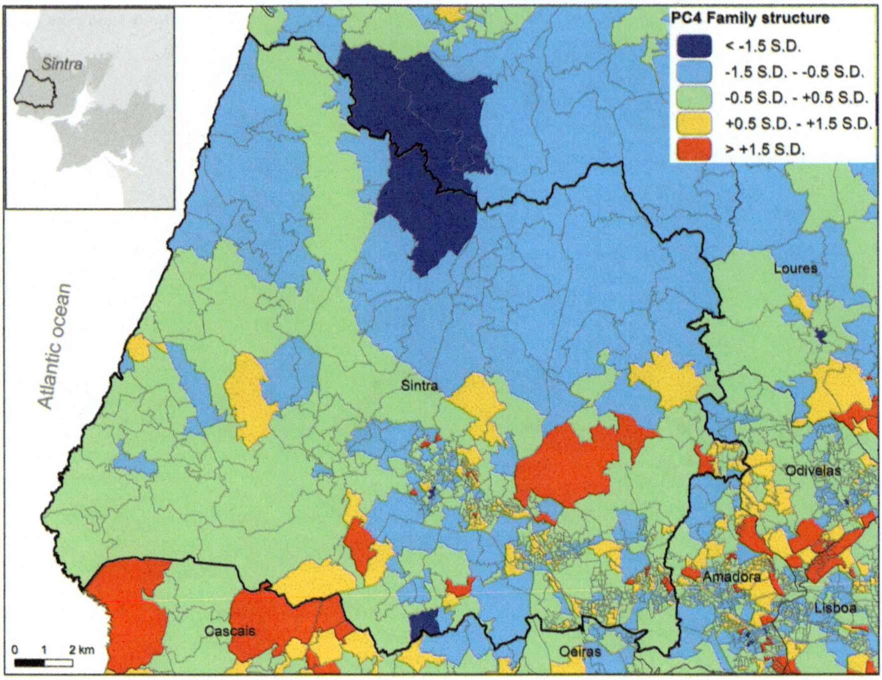

Fig. 8 Cartographic expression of the PC scores in regard to the family structure (PC4 of criticality) in the Sintra municipality

Table 6 Final set of variables used in the support capability assessment, after redundancy elimination and analysis of robustness

Code	Variable	Communalities
Hotel housing	Coverage by hotels with capacity for temporary shelter	0.904
Fire sta	Coverage by fire stations	0.632
Pharm	Coverage by pharmacies	0.756
Road nodes 34	Coverage by the main road network	0.491
Hospital	Coverage by hospitals	0.587
Health centre	Coverage by health centres	0.653
Gas sta	Coverage by car gas and power stations	0.630
Police	Coverage by police stations	0.691
Grocery	Coverage by grocery stores	0.583

is important to focus on the detailed level of analysis made possible by the small dimension of the statistical blocks.

Table 7 Rotated component matrix for the assessment of support capability

	Principal components (PCs)	
	1	2
Coverage by pharmacies	0.861	0.124
Coverage by police stations	0.810	0.186
Coverage by car gas and power stations	0.793	0.024
Coverage by fire stations	0.769	0.203
Coverage by grocery stores	0.762	0.053
Coverage by health centres	0.759	0.279
Coverage by the main road network	0.657	0.242
Coverage by hospitals	0.565	0.518
Coverage by hotels with capacity for temporary shelter	−0.051	−0.950
Cardinality	–	–
% of the total variance explained	54.741	11.135

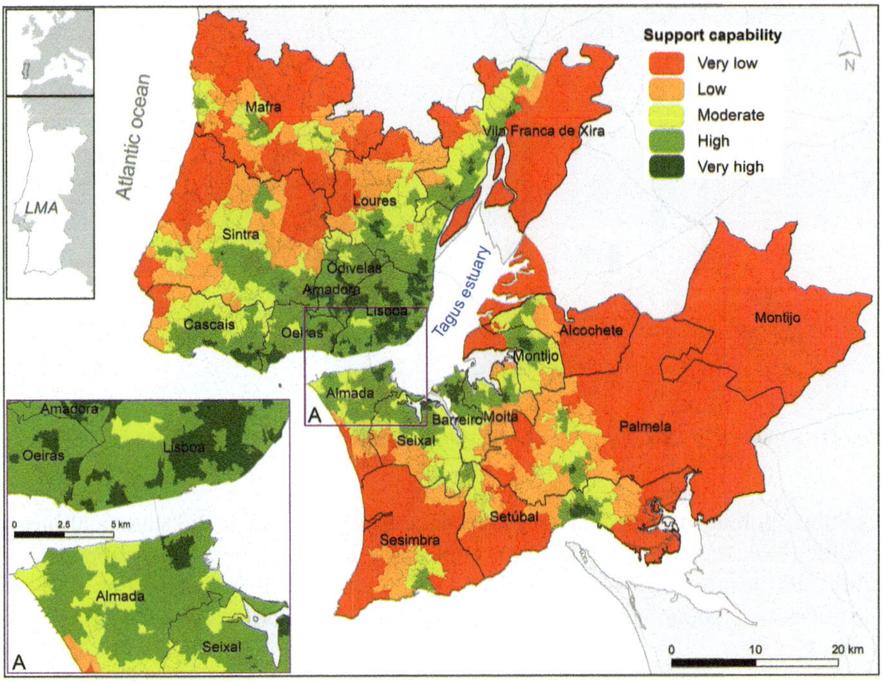

Fig. 9 Support capability in the Lisbon Metropolitan Area

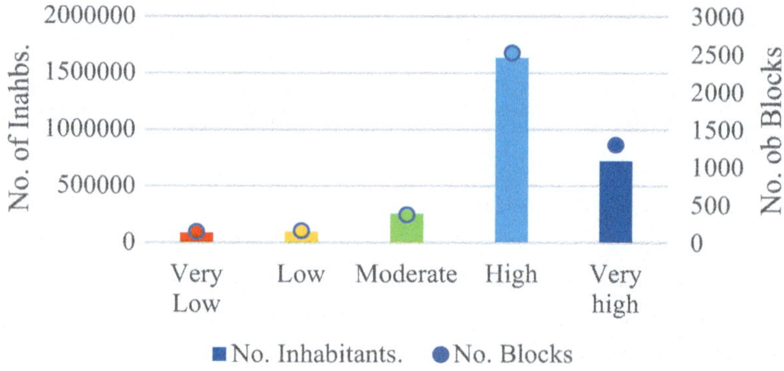

Fig. 10 Number of inhabitants and statistical blocks by class of support capability in the LMA

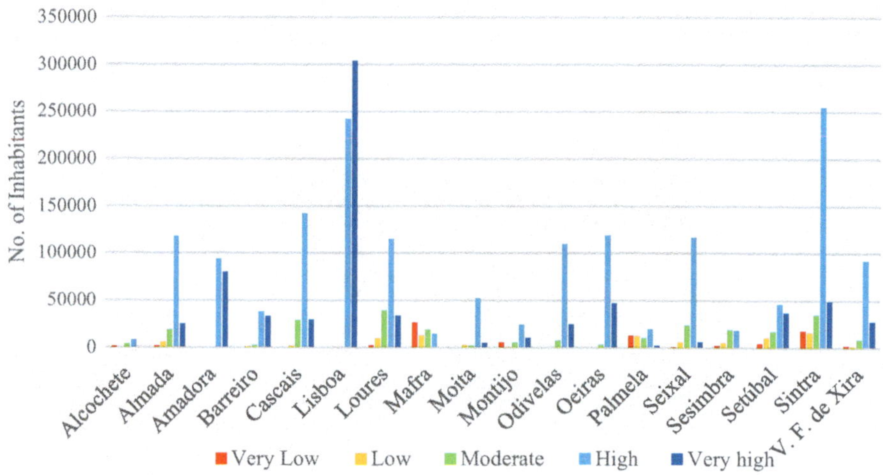

Fig. 11 Summary of inhabitants by class of support capability in the LMA municipalities

Public policies related to risk governance need to consider the statistical blocks resulting from the overlay of the least supported areas (low and very low support capability) and the most critical ones (high and very high criticality) because they represent the hotspots of social vulnerability in the Lisbon Metropolitan Area (Fig. 14).

Table 8 Summary of the number of inhabitants and statistical blocks by class of support capability in the 18 municipalities of the LMA

Municipalities	Very low		Low		Moderate		High		Very high		Totals	
	No. blocks	No. inhab.	No. blocks	No. inhab.	No. blocks	No. inhab.	No. blocks	No. inhab.	No. blocks	No. inhab.	Blocks	Inhab.
Alcochete	5	2875	1	492	7	5085	14	9117	0	0	27	17,569
Almada	7	2679	18	6879	38	19,635	209	119,028	46	25,809	318	174,030
Amadora	0	0	0	0	0	0	141	94,440	138	80,696	279	175,136
Barreiro	0	0	4	2408	7	3691	66	38,822	57	33,843	134	78,764
Cascais	1	456	4	2758	37	30,118	226	142,896	52	30,251	320	206,479
Lisboa	0	0	0	0	1	516	453	242,797	600	304,420	1054	547,733
Loures	6	3714	17	10,957	55	40,373	163	115,358	54	34,652	295	205,054
Mafra	40	27,394	19	13,863	27	20,127	26	15,301	0	0	112	76,685
Moita	0	0	6	3975	5	3105	80	52,883	10	6066	101	66,029
Montijo	9	6566	2	1547	11	6616	34	25,157	19	11,336	75	51,222
Odivelas	0	0	0	0	9	8635	147	110,467	42	25,447	198	144,549
Oeiras	0	0	0	0	5	4269	184	119,875	84	47,976	273	172,120
Palmela	21	13,872	22	13,692	18	11,617	30	20,448	5	3202	96	62,831
Seixal	3	1793	12	7187	36	24,793	178	117,596	12	6900	241	158,269
Sesimbra	11	3489	11	6561	29	20,146	36	19,304	0	0	87	49,500
Setúbal	11	5806	17	12,078	27	18,545	69	46,941	68	37,815	192	121,185
Sintra	30	19,115	26	17,232	49	36,136	344	255,555	79	49,797	528	377,835
VF Xira	5	3086	4	2729	15	9816	125	92,865	42	28,390	191	136,886
Total	149	90,845	163	102,358	376	263,223	2525	1,638,850	1308	726,600	4521	2,821,876

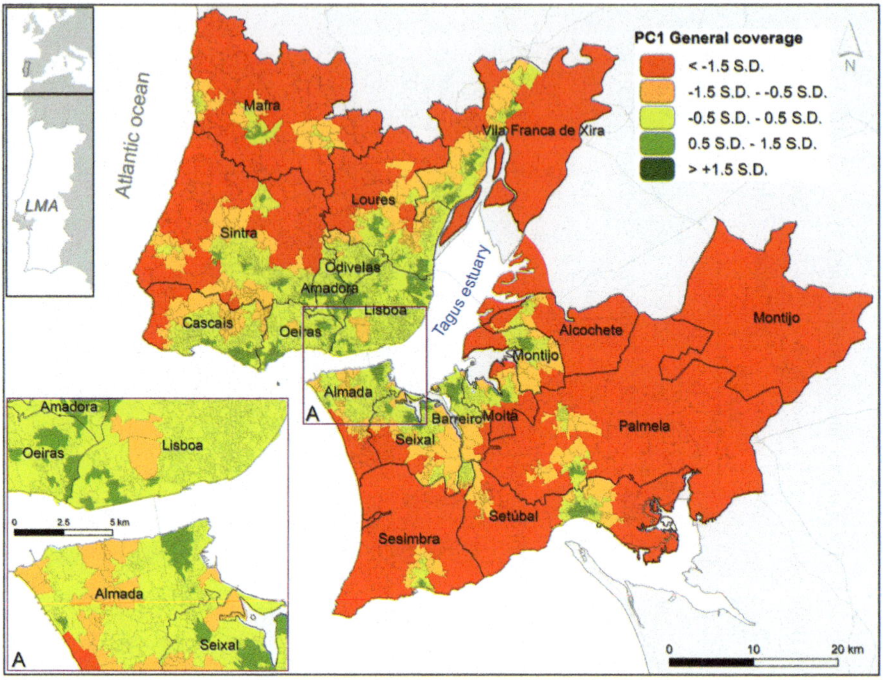

Fig. 12 Cartographic expression of the PC scores in regard to the general equipment and services coverage (PC1 of support capability) in the LMA

5 Conclusions

Social vulnerability is a key indicator for risk governance, involving the processes and impacts arising from events of natural, technological, or environmental origin [10, 14]. Several studies suggest that the levels of vulnerability of individuals and communities explain (in certain geographical and socioeconomic contexts) the impacts observed in databases, as much as the levels of susceptibility and exposure to hazard processes.

In this study, we applied a methodology for assessing social vulnerability in the Lisbon Metropolitan Area (LMA), based on principal component analysis, from 2011 Census data. The territorial unit of analysis is the statistical section, making up 4521 units, consisting of a very detailed-level, large-scale analysis for the entire LMA.

Starting from an initial set of 43 variables in the domains of age, gender, employment, educational qualifications, housing conditions, and mobility, the final model for criticality integrates 12 variables, extracting four principal components (PC), interpreted as follows: employment and qualifications (PC1), which explains 32.5% of the total variance; age, gender, and ageing urban context (PC2), 22.5% of the total variance; housing conditions (PC3), 10.3% of the total variance; and family

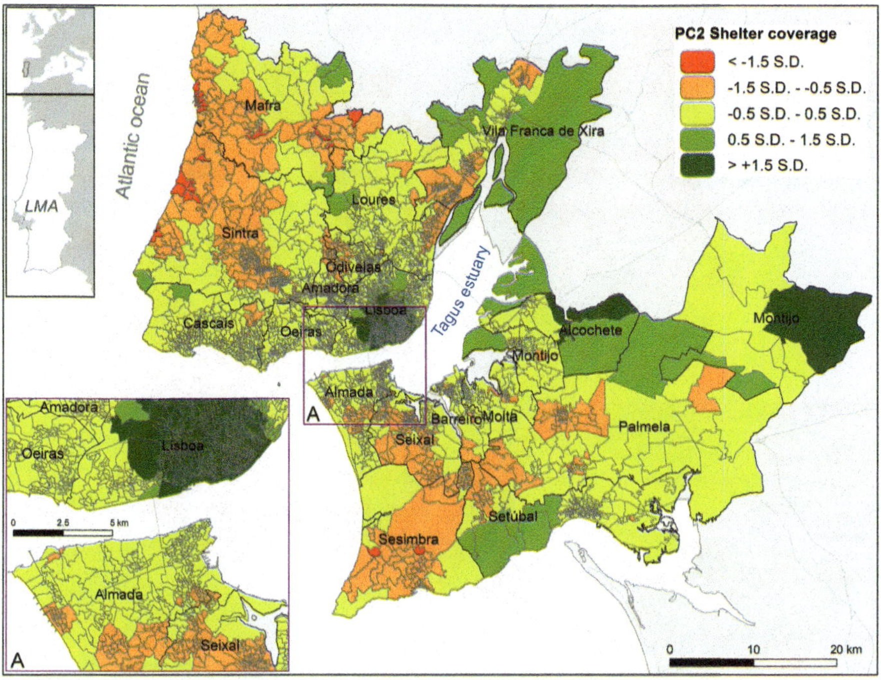

Fig. 13 Cartographic expression of the PC scores in regard to the shelter coverage (PC2 of support capability) in the LMA

structure (PC4), 8.4% of the total variance. The sum of the scores of each principal component provides a final index of criticality that allows the identification of the most vulnerable neighbourhoods and urban centres.

For the support capability assessment, an initial set of 9 variables were considered, and the resulting PCs express the general coverage by most of the considered services and equipment (PC1) and the particular coverage by the equipment with the ability to provide temporary shelter (PC2).

The analysis of the individual mapping of the scores of each component provides an understanding of the most active dimensions or drivers of criticality and support capability in each statistical section.

Social vulnerability studies are applied at two levels of public policy action: in supporting emergency civil protection planning for the phases of imminence, occurrence, and post-disaster recovery; in medium and long-term risk management planning, identifying, and understanding the drivers that explain the propensity of individuals and communities to loss and the degree of difficulty in recovery. Both levels translate into the definition of intra- and inter-municipal resource allocation priorities that promote increased resilience to various types of risks.

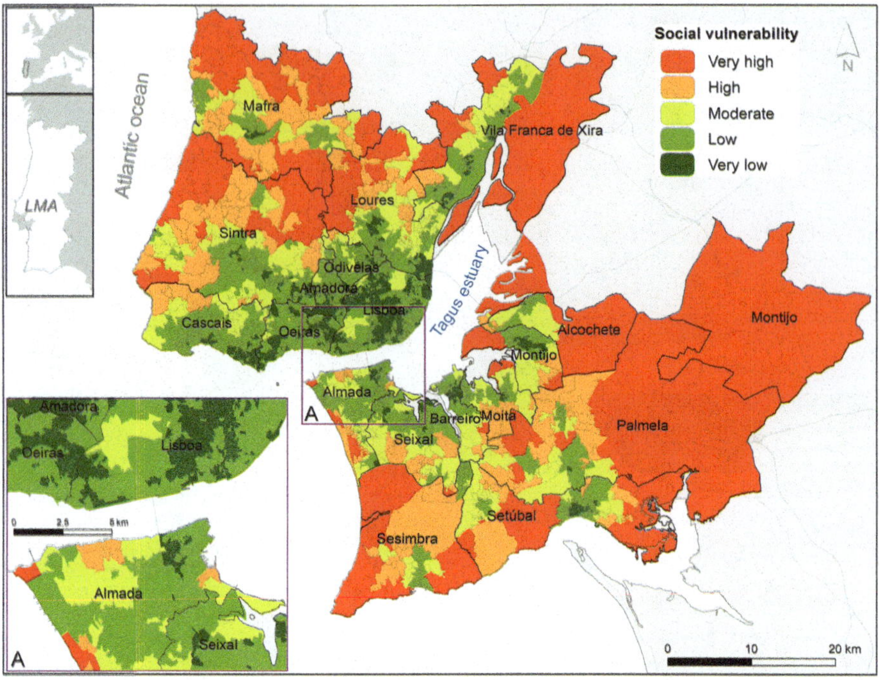

Fig. 14 Social vulnerability in the Lisbon Metropolitan Area

Acknowledgements The project 'MIT-RSC—Multi-risk Interactions Towards Resilient and Sustainable Cities' (MIT-EXPL/CS/0018/2019) leading to this work is co-financed by the ERDF—European Regional Development Fund through the Operational Program for Competitiveness and Internationalisation—COMPETE 2020, the North Portugal Regional Operational Program NORTE 2020, and by the Portuguese Foundation for Science and Technology—GCT under the MIT Portugal Program at the 2019 PT call for Exploratory Proposals in 'Sustainable Cities'. RISKCOAST is funded by the Interreg Sudoe Programme (SOE3/P4/E0868).

References

1. Chen W, Cutter SL, Emrich CT, Shi P (2013) Measuring social vulnerability to natural hazards in the yangtze river delta region, China. Int J Disaster Risk Sci 4(4):169–181. https://doi.org/10.1007/s13753-013-0018-6
2. Cutter SL, Boruff BJ, Shirley WL (2003) Social vulnerability to environmental hazards. Soc Sci Q 84(2):242–261. https://doi.org/10.1111/1540-6237.8402002
3. Cutter SL, Finch C (2008) Temporal and spatial changes in social vulnerability to natural hazards. Proc Natl Acad Sci 105(7):2301–2306. https://doi.org/10.1073/pnas.0710375105
4. Eidsvig UMK, McLean A, Vangelsten BV, Kalsnes B, Ciurean RL, Argyroudis S, Winter MG et al (2014) Assessment of socioeconomic vulnerability to landslides using an indicator-based approach: methodology and case studies. Bull Eng Geol Env 73(2):307–324. https://doi.org/10.1007/s10064-014-0571-2

5. Frigerio I, Carnelli F, Cabinio M, De Amicis M (2018) Spatiotemporal pattern of social vulnerability in Italy. Int J Disaster Risk Sci 9(2):249–262. https://doi.org/10.1007/s13753-018-0168-7
6. IPCC-Working Group II (2022) Climate change 2022-6th assessment report. Impacts, adaptation and vulnerability. summary for policy makers. Working Group II. Intergovernmental Panel on Climate Change
7. Jamshed A, Rana IA, Mirza UM, Birkmann J (2019) Assessing relationship between vulnerability and capacity: an empirical study on rural flooding in Pakistan. Int J Disaster Risk Reduct 36:101109. https://doi.org/10.1016/J.IJDRR.2019.101109
8. Mendes JM (2009) Social vulnerability indexes as planning tools: beyond the preparedness paradigm. J Risk Res 12(1):43–58. https://doi.org/10.1080/13669870802447962
9. Mendes JM, Tavares AO, Cunha L, Freiria S (2011) A vulnerabilidade social aos perigos naturais e tecnológicos em Portugal [Social vulnerability to natural and technological hazards in Portugal]. Revista Crítica de Ciências Sociais 93:95–128
10. Mendes JM, Tavares AO, Santos PP (2019) Social vulnerability and local level assessments: a new approach for planning. Int J Disaster Resil Built Environ 11(1):15–43. https://doi.org/10.1108/IJDRBE-10-2019-0069
11. Ogie RI, Pradhan B (2019) Natural hazards and social vulnerability of place: the strength-based approach applied to wollongong, Australia. Int J Disaster Risk Sci 10(3):404–420. https://doi.org/10.1007/s13753-019-0224-y
12. Rufat S, Tate E, Burton CG, Maroof AS (2015) Social vulnerability to floods: review of case studies and implications for measurement. Int J Disaster Risk Reduct 14:470–486. https://doi.org/10.1016/j.ijdrr.2015.09.013
13. Rufat S, Tate E, Emrich CT, Antolini F (2019) How valid are social vulnerability models? Ann Am Assoc Geogr 109(4):1131–1153. https://doi.org/10.1080/24694452.2018.1535887
14. Santos PP, Tavares AO, Freire P, Rilo A (2018) Estuarine flooding in urban areas: enhancing vulnerability assessment. Nat Hazards 93:77–95. https://doi.org/10.1007/s11069-017-3067-0
15. Santos PP, Zêzere JL, Pereira S, Rocha J, Tavares AO (2022) A novel approach to measuring spatiotemporal changes in social vulnerability at the local level in Portugal. Int J Disaster Risk Sci 13(6):842–861. https://doi.org/10.1007/s13753-022-00455-w
16. UNDRR (2019) Global assessment report 2019. United nations office for disaster risk reduction. Geneva, Switzerland
17. UNDRR (2022) UNDRR strategic framework 2022–2025. United nations office for disaster risk reduction. Geneva, Switzerland
18. Windfeld EJ, Ford JD, Berrang-Ford L, McDowell G (2019) How do community-level climate change vulnerability assessments treat future vulnerability and integrate diverse datasets? A review of the literature. Environ Rev 27(4):427–434. https://doi.org/10.1139/er-2018-0102

Chapter 4
Flood Risk Assessment in the Lisbon Metropolitan Area

Pedro Pinto Santos, Maria Xofi, José Carlos Domingues, and Tiago Miguel Ferreira

Abstract Flood processes are one of the most challenging to risk assessment and management. In many situations, peak flows are generated kilometers away from the places where inundation is observed. Scale in flood risk assessments is a fundamental factor when estimating hazard, exposure, and vulnerability. Municipal, civil parish, and building-level information are used to construct flood risk indexes and profiles. It is observed that, depending on the scale at which it is represented, the same root information provides distinct insights into flood risk expression in the Lisbon Metropolitan Area. When compared with the Flood Directive critical areas, the results show they are mostly consistent with the results at the different scales, identifying the same hotspots of flood risk (in the Loures, V. F. Xira, and Setúbal municipalities) as those selected during the Directive's implementation. Flood loss reduction implies the involvement of distinct risk practitioners and decision-makers, acting at distinct scales and sectors related to risk governance. Interconnections between flood risk components and between flood processes and other potential cascading processes are still insufficiently known and require the priority of society.

Keywords Flood risk assessment · Flood hazard · Flood risk index · Flood vulnerability · Metropolitan scale

---

P. P. Santos (✉)
Centre for Geographical Studies, Institute of Geography and Spatial Planning (IGOT), LA TERRA, University of Lisbon, Lisbon, Portugal
e-mail: pmpsantos@campus.ul.pt

M. Xofi
Faculty of Engineering Technology, University of Twente, Enschede, The Netherlands
e-mail: mariaxofi@hotmail.com

M. Xofi · J. C. Domingues
Institute for Sustainability and Innovation in Structural Engineering (ISISE), Department of Civil Engineering, University of Minho, Guimarães, Portugal

T. M. Ferreira
School of Engineering, College of Arts, Technology and Environment (CATE), University of the West of England (UWE Bristol), Bristol, UK
e-mail: Tiago.Ferreira@uwe.ac.uk

T. M. Ferreira (ed.), *Multi-risk Interactions Towards Resilient and Sustainable Cities*, Advances in Sustainability Science and Technology,
https://doi.org/10.1007/978-981-99-0745-8_4

1 Introduction

Small and frequent flood disasters, along with catastrophic and rare events, continue to affect our environmental and socioeconomic living, despite the advances in spatial planning, engineering solutions, monitoring, and early warning systems [29]. As in many other processes, in regard to floods, nature explains only partly the degree of damage observed [22].

Extreme hydrological phenomena, especially river flooding, are considered one of the most important disasters induced by natural processes in Europe in terms of economic losses [6, 19]. Moreover, it is increasingly accepted that global warming modifies the hydrological cycle and increases the occurrence and frequency of flood events in several regions of this continent and worldwide [6].

In 2007, the European Union assumed a new and uniform framework for the assessment and management of flood risks for all its member states. This framework, defined in Directive 2007/60/EC of the European Parliament and Council of 23 October, and transposed into the Portuguese national law by Decree-Law No. 115/2010 of 22 October, introduces a set of challenges to which the scientific community cannot and must not be oblivious.

The approved framework highlights the importance of flood risk mapping, from which the respective management plans (Flood Risk Management Plans—FRMP) are defined. The challenges posed by the development of these planning instruments—the choice of non-structural measures, the possibility of controlled flooding, the scale of analysis adopted, and the linkage with land use plans and civil protection emergency plans—demonstrate the relevance of flood risk studies that bring detailed and comprehensive knowledge, allowing a better connection between the risk assessment and management processes, framed in risk governance models [4].

Conceptual models of risk usually consider the hazard, exposure, and vulnerability components ([7, 24]). In the past, although implicit, exposure was not always considered an autonomous risk component [21, 30, 32]. In flood risk studies within the climate change community until recently, exposure was considered in risk models as a subcomponent of vulnerability [5, 25, 33]. Some risk models incorporate the societal and institutional dimensions of the capacity to cope and adapt to threats [14], for instance).

For the sake of homogeneous applicability in the entire European Union, the vulnerability concept is absent in the Flood Directive. In fact, although being an invaluable concept in risk analysis (IPCC, 2022, WG II), a recent study in Northern Portugal demonstrated that flood damage databases are spatially more correlated with areas of high hazard and exposure than to areas of high vulnerability [26].

Apart from the conceptual dynamics in flood risk analysis, the scale of assessments was and still is an important aspect to consider. Different scales require different data and methods, while the intents of the produced analysis also influence the methodologies to adopt [20]. Consistency and accuracy of results are the more concerning issues when flood risk is compared across scales. Vulnerability and flood protection representation is a critical aspect of national and global flood risk analysis [20]. Global datasets provided by global Earth observation systems and more powerful computational capacity seem to be shortening, or concealing, some of the constraints imposed by scale, thus allowing for increasingly detailed flood risk analysis to be produced to vaster areas, i.e., smaller scales [2, 12, 15, 31].

In the context of the MIT-RSC project, flood risk was evaluated at distinct scales of analysis aiming to produce distinct risk products. The effect of scale in flood risk assessment was considered a key factor in the representation of results, as they are required by the several stakeholders involved in flood risk governance, at the distinct geographical, administrative, and sectorial levels.

In this chapter, a municipal, a civil parish, and a building-level assessment of flood risk was carried out using scale-specific input data expressing hazard, exposure, physical vulnerability of buildings, and social vulnerability of the resident population. The aim is to contribute to this research area by providing insights into the constraints, particularities, and significance of spatial scale in the design of flood risk indexes.

2 Study Area

The Lisbon Metropolitan Area (LMA) is a statistical and administrative entity that comprises an area of 3,105 km^2 and 2,813,000 inhabitants (27% of the Portuguese population). The LMA is geographically highly contrasting: dense urban areas near the capital city of Lisbon and along the main road and rail network, contrast with vast areas corresponding to legally protected areas (Tagus and Sado estuaries, Arrábida and Sintra's mountains) and forestry and farmland areas mainly in the southern municipalities of the LMA.

The work developed under the implementation of the Flood Directive identified four areas of significant potential flood risk (ASPFR) in the Lisbon Metropolitan Area (LMA). The 2nd implementation cycle adds a few more, although data is still not available. One of the ASPFR is subject to slow onset flooding, and three areas are subject to flash flooding (Fig. 1):

- Abrantes-Tagus estuary: fluvial, slow onset flooding affecting the Tagus river and part of its estuary covering, in the LMA, the municipality of V.F. Xira;
- Loures-Odivelas: fluvial flash flooding in the Trancão river, a tributary of the Tagus river, covering the homonymous municipalities;
- Torres Vedras: fluvial flash flooding in the Pequeno river (a small tributary in the Mafra municipality, belonging to the Sizandro river basin);
- Setúbal: fluvial flash flooding in the Livramento stream, affecting the Setúbal municipality.

When addressing hazards such as floods, which particularly affect floodplains and some urban areas, the scale of representation of risk components (hazard, exposure, and vulnerability) is of major importance. Figure 2a and b highlights that representing exposed elements as the resident population at an intermediate level like the civil

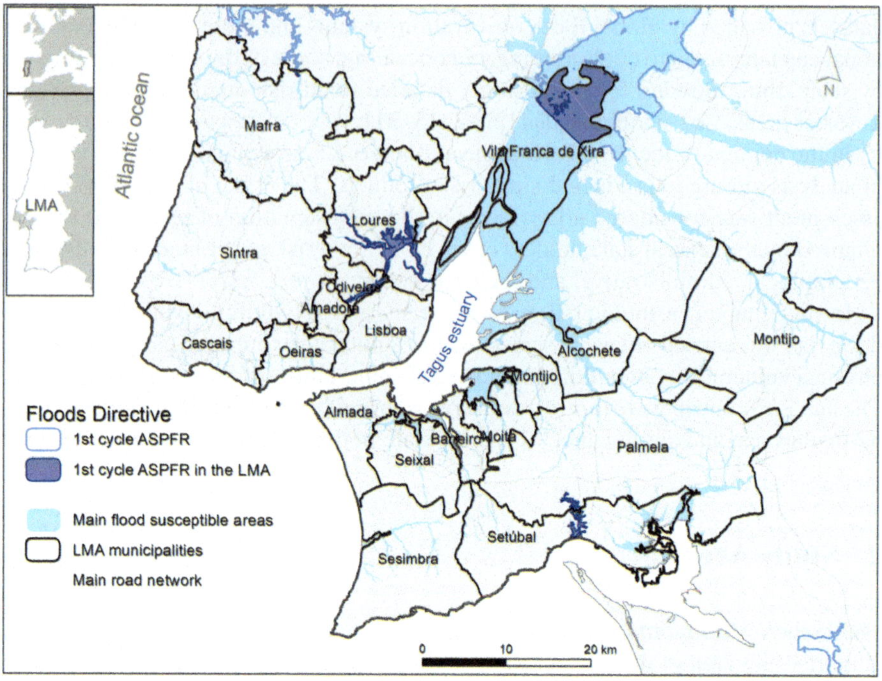

Fig. 1 Municipalities of the Lisbon Metropolitan area and the flood directive's areas of significant potential flood risk

parish conceals the local hotspot of urban concentration that is only visible at the census statistical block level.

3 Methodology

3.1 *Hazard Assessment*

3.1.1 Municipal Level

Flood hazard was evaluated at the municipal level (*Mun_H*) considering five parameters: maximum flood event recorded (H1), which represents the maximum historical event registered in the last 150 years considering the DISASTER database [34], frequent flood event (H2), representing the total amount of events recorded in the database, and occurred in a given municipality independently of the degree of loss (very often, the cumulative effects of a frequent event can be more impacting in the long term than low probability/high consequence events); annual exceedance probability of the maximum flood event recorded in all the 18 municipalities (H3); annual

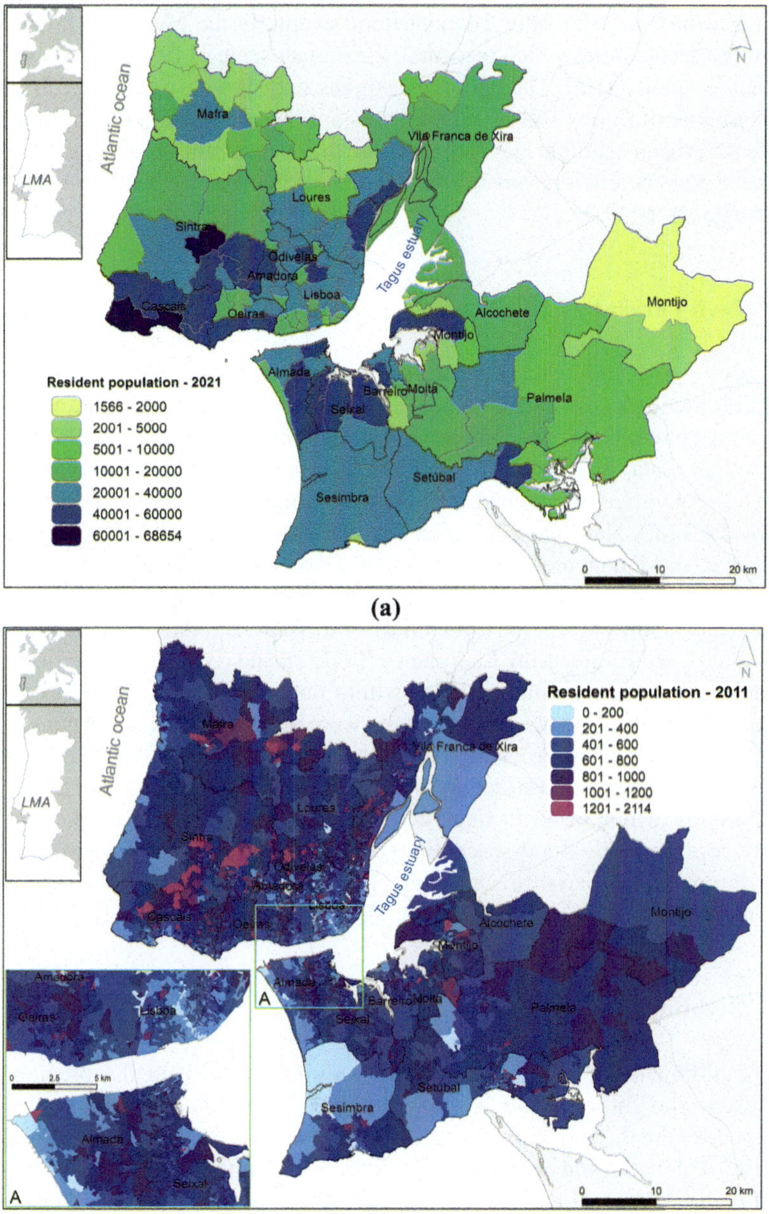

Fig. 2 Resident population expressed at the civil parish (**a**) and Census statistical block (**b**)

exceedance probability of the frequent flood events in the LMA (H4); and spatial scale of the assessment, which measures the impact scale of flood hazard within the entire municipality (H5). Data sources express both the susceptibility, magnitude, and recurrence of floods. All five parameters are represented at the municipal level. A weighted mean valuing H5 with 40% of the weight and assigning 15% to the remaining four parameters was calculated and normalized by the min–max method to the range [0, 1].

3.1.2 Civil Parish Level

At the civil parish level, flood hazard (CP_{FH}) was assessed considering the flood-susceptible areas identified using a more straightforward approach based on historical data, geomorphological interpretation of topographical maps, and satellite imagery.

Historical data consisted of available documentation from previous works, namely: (i) the areas threatened by floods delimited within the Regional Framework of the LMA National Ecological Reserve; (ii) the floodable areas defined in the Flood Risk Management Plans, made available by the Portuguese Environment Agency [3], (iii) the delimitation of the flood associated with the 1979 Tagus River flood and the delimitation of the centennial flood in the Sado estuary area, produced by the National Civil Engineering Laboratory (LNEC); and (iv) a set of flood-threatened areas delineated at the municipal level, within the scope of the National Ecological Reserve delimitation. Ponds and reservoirs were also identified as permanent water bodies. As regards flood processes, slow onset flood situations were distinguished, associated with the main watercourses in the region (e.g., Tagus River), from cases of flash floods, which occur in small and medium hydrographic basins.

After compiling the flood-susceptible polygons, a simple proportion of that area within the entire civil parish area was calculated and expressed in %, later normalized by the min–max method to the range [0, 1].

3.1.3 Building Level

At the building level, the approach differs significantly from the previous ones. Flood extent, height, and velocity were obtained from 2-D hydraulic modeling [3] performed within the implementation of the Flood Directive. Results were validated with historical marks and georeferencing from flood damage databases. Data is available from the APA's webGIS for the four ASPFR existing in the LMA and presented above.

The buildings' database (BGE) is sourced from the Statistics Portugal database regarding the 2011 population Census. Each building was assigned a flood depth and velocity, considering the mean value in a search radius of 10 m from the point feature.

Clusters of buildings were obtained by running first a Two-Step Cluster Analysis in SPSS©, testing 2–10 clustering solutions with Schwarz's Bayesian Criterion

(BIC). The statistical individuals are the 2766 buildings located in the four ASPFR. The variables considered were the physical vulnerability scores, flood depths, and velocities per building.

3.2 Exposure Assessment

The methodology for representing flood exposure at the municipal, civil parish, and building level makes use of the same input data—the Georeferenced Buildings Database (BGE) mentioned above. In fact, this detailed layer of geographical information exists uniformly for the entire LMA.

The exposure module identifies and characterizes the exposed elements in each polygonal unit of analysis (municipality and civil parish), focusing on the residential buildings—approx. 450,000 points of the BGE—and resident population, taken from the Geographic Base for Information Referencing (BGRI, both obtained from Statistics Portugal). The dasymetric distribution of the resident population by buildings with a total or partial residential function was performed [8]. The refinement of exposure to flooding was done by quantifying the number of buildings and respective inhabitants within flood-susceptible areas (the same ones used to define the flood hazard parameter H5), by municipality and by civil parish.

At the municipal level, a transformation to the range [0, 1] was done accounting for the maximum (1813 buildings in the Lisboa municipality) and minimum (5 buildings in the Sesimbra municipality) number of buildings exposed to flooding. This procedure defines municipal buildings' exposure. A similar approach is applied to the resident population (a maximum of 17,675 in Lisboa, and a minimum of 4 in Sesimbra), resulting in a transformation of population exposure to flooding between 0 and 1. A weight of 0.25 was assigned to buildings and 0.75 to population, of which sum results in the final exposure, also ranging [0, 1].

At the civil parish level, the same method was followed. In order to avoid the absorbing effect of zero on the multiplication, scores of 0 were replaced by 0.0001 except in the cases where there's no flood susceptibility or no buildings within susceptible areas.

At the building level, exposure was assessed only in ASPFR polygons. Inside each of the four ASPFR polygons, the number of residents (the one resulting from the dasymetric distribution) and the number of buildings were summed and later converted to the range [0, 1].

3.3 Buildings' Physical Vulnerability Assessment

The assessment of the physical vulnerability of buildings (*PhyV*) uses as input data the variables' values collected during the 2011 Census operation. As presented, this data is part of the BGE database which comprises 14 variables describing the buildings' characteristics. From the set of BGE's 14 variables, 6 were used to assess the physical vulnerability of buildings (Table 1), to which one other (P5) was added from geological and soil characteristics.

The table describes the theoretical and empirical basis that supports both the selection of parameters and their role in increasing or decreasing vulnerability (Table 1).

Table 1 Parameters adopted to represent the physical vulnerability of buildings, their rationale, and supporting academic literature

Parameters		Argumentation/Rationale	Sources
P1	Period of construction	The building age is linked to the building condition (similar to parameter P7). It is considered a common indicator revealing different construction techniques and materials. Therefore, it is ranked as less important to flood vulnerability compared to other parameters	Agliata et al. [1], Pereira et al. [10], Leal et al. [11] and Kappes et al. [23]
P2	No. stories [a]	1. The taller the building, the deeper the foundation and the less prone to be damaged by flooding 2. The taller the building, the more vulnerable to flooding and susceptible to differential settlements considering poor ground conditions and/or shallow footings	1. Agliata et al. [1], Kappes et al. [10], Pereira et al. [23] 2. Miranda and Ferreira [18], Stepchenson and D'Ayala [28]
P3	Material-Ext. cladding	The absorption of the cladding material directly affects the building fabric resulting in a direct impact on the building's susceptibility	Common knowledge and Leal et al. 11] and Kappes et al. [10]
P4	Material-Str. system	Depending the type of external cladding material and its absorption, it will affect directly the building fabric. Hence, it is directly related to the building's susceptibility to flooding impact	Common knowledge and Leal et al. 11] and Kappes et al. [10]
P5	Soil classification	The type of soil has a direct impact on the foundations and scour due to flooding	Common knowledge/empirically
P6	Building exposure	This is related to the position of the building and its exposure. It is considered less significant to flooding given the level of uncertainty compared to other indicators	Agliata et al. [1], Kappes et al. 10]

(continued)

Table 1 (continued)

Parameters		Argumentation/Rationale	Sources
P7	Repair Req/condition[b]	Existing condition and maintenance status of a building is linked directly with the building fabric material, as per parameter P3. Hence, parameter P7 is also considered a key influence factor	Agliata et al. [1], Kappes et al. [10], Silva and Pereira 27, Stepchenson and D'Ayala [28]

[a] Most debated indicator as experts interpret and use it in different ways depending on what building feature and behavior they intend to describe with it [1]
[b] Most used indicator reflecting the capacity of construction to resist flood impact

It also provides an overview of the weighting (i.e., importance factor) of parameters that contribute to the building's potential (i.e., resistance capacity of the building) to withstand a flood impact.

The vulnerability assessment to flooding considers the period of construction (P1), the number of stories (P2), the material of the external cladding (P3), the material of the structural system (P4), soil/lithological substrate (P5), building exposure (P6), and building condition (P7) (see Fig. 3). Each parameter is then evaluated using four vulnerability classes A to D, where A represents the least vulnerable condition, and D represents the most vulnerable one.

The physical vulnerability of buildings to flooding (*PhyV*) method is based on the estimation of an index for each building as the weighted sum of a set of seven parameters, as listed above, using Eq. 1. For easier interpretation of results and integration with hazard and exposure results, *PhyV* scores were normalized to the

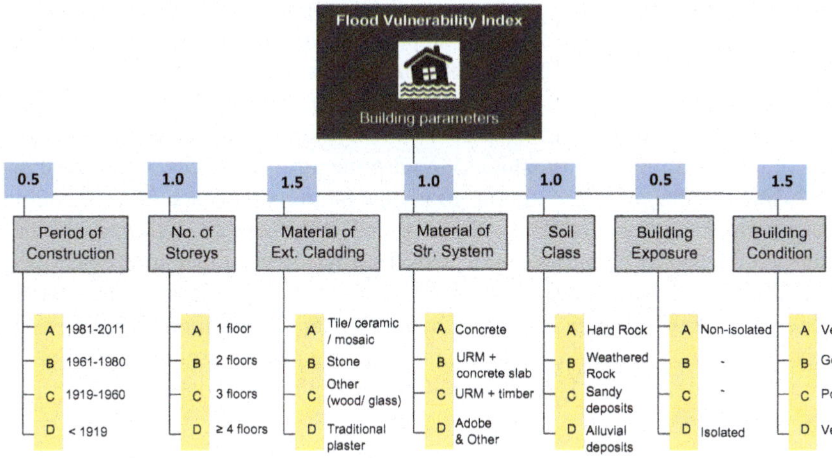

Fig. 3 Parameters, properties, and respective weights used in the assessment of buildings' physical vulnerability to flooding

range [0, 1] using the min–max method.

$$PhyV = \sum_{p=1}^{7} (C_p \times W_p) \tag{1}$$

where C_p is the score assigned through classes A to D in each parameter p, and W_p is the weight assigned in each parameter p.

3.4 Social Vulnerability

Social vulnerability ($SocV$) was evaluated for the entire LMA using the Census statistical block level and is the result of the product of criticality (Cr) and support capability (SC) [17], as formulated in Eq. 2.

$$SocV = Cr \times (1 - SC) \tag{2}$$

In the LMA there are 4521 statistical blocks. The proposed method incorporates not only the individual characteristics of the population and risk groups as it also considers the territorial context in which they are supported, i.e., the public and private equipment, infrastructure, and services that might play a role in attenuating losses and enhancing recovery [16].

Criticality (Cr) expresses the characteristics of individuals that make them prone to loss, considering their age, socioeconomic condition, health and housing conditions, social assistance, mobility, educational level, and employment. Support capability (SC) expresses the set of systems, networks, public and private infrastructures, and collective equipment aimed at supporting communities and their activities, which—in the eminence or occurrence of a dangerous process—make it possible to reinforce a community's capacity to mitigate and/or recover from a hazardous event. Common dimensions covered are the economic dynamism, the coverage by social equipment (for example, health centers), civil protection resources, and public and private businesses that provide essential goods and mobility. For each of these components, a principal component analysis is performed, and interactively, the most robust and interpretable set of variables is selected, from which principal component (PC) scores are extracted. Those scores are summed according to the weights provided by the percentage of variance explained by each principal component. Prior to the application of Eq. 2, the final criticality and support capability scores were linearly transformed to the range [0, 1]. For a detailed description of variables and methods, please refer to Chap. 3.

3.5 Risk Analysis

At the municipal level, municipal flood risk (Mun_{FR}) was calculated as follows (Eq. 3):

$$Mun_{FR} = Mun_{FH}^{1/3} \times Mun_{E}^{1/3} \times Mun_{PhyV}^{1/3} \tag{3}$$

where Mun_{FH} is the municipal flood hazard, Mun_E is the municipal exposure, and Mun_{PhyV} is the municipal physical vulnerability of buildings.

At the civil parish level, flood risk was calculated using the physical vulnerability of buildings on one side and the criticality (*Cr*) of social vulnerability on the other. Civil parish flood risk using criticality is calculated as follows (Eq. 4):

$$CP_{FR\_Cr} = CP_{FH}^{1/3} \times CP_{E}^{1/3} \times CP_{Cr}^{1/3} \tag{4}$$

Civil parish flood risk using physical vulnerability (CP_{FR_PhyV}) is calculated as follows (Eq. 5):

$$CP_{FR\_PhyV} = CP_{FH}^{1/3} \times CP_{E}^{1/3} \times CP_{PhyV}^{1/3} \tag{5}$$

Like in the municipal level analysis, all scores expressing hazard, exposure, and vulnerability are normalized to the range [0, 1] before the INFORM alike formulation is applied by calculating the product of the power of each component to 1/3.

Building-level flood risk was assessed using a cluster analysis that provides flood risk profiles. For each of the 2766 buildings located in the four ASPFR, a cluster membership is calculated that considers the hazard (flood depth and velocity) and the physical vulnerability (*PhyV*) scores as described in Sect. 2.3.

4 Results

4.1 Municipal Level

Flood hazard in the LMA is significantly conditioned by the Tagus river floodplain, the Tagus estuary, and the small watersheds that drain to the Tagus, directly to the Atlantic ocean or to the municipalities outside the LMA. The maximum flood event (H1) in each municipality differs, although two events stand out: the flash floods of November 1967 and November 1983, which mostly impacted the municipalities of V.F. Xira, Loures, Odivelas, and Oeiras (Fig. 4). Adding to these municipalities, the higher scores in regard to the frequent events (H2) include the district capitals of Lisbon and Setúbal, as well as Almada and Mafra. Parameter H3 assumes the major of the maximum events could, theoretically, have affected evenly any of the

18 municipalities. For this reason, an equal score was assigned to all municipalities with a low probability event [13]. The same principle was applied to the probability of the frequent event in parameter H4. Finally, H5 highlights the municipalities of V.F. Xira, Odivelas, and Loures due to the highest proportion of flood-susceptible areas in the corresponding territory (60.4, 11.5, and 11.1%, respectively). These four municipalities and Oeiras are characterized by a very high and high flood hazard (Fig. 4) as a result of a high spatial propensity and historical record of floods, either with a high magnitude or a high degree of loss. Exposure is higher in the most urbanized municipalities, and it is linked to the occupation of the small streams valley floors, particularly in Lisbon, Odivelas, and Setúbal (Table 2). Globally, 2.5% of the LMA residents live in flood-susceptible areas. When exposure is combined with the physical vulnerability of buildings, it becomes more evident where the most critical contexts for potential flood losses are located. The mean PV in the AML is 0.35, with maximum average values of 0.50 in Lisbon and 0.44 in Setúbal. Nevertheless, a detailed analysis of PV needs to consider each parameter individually; this is because, for example, the general condition of the buildings (Table 1)—as it was recorded in the Census—may not be concerning, but other parameters like lithological substrate, the material of the external cladding, and structural system do act as stronger drivers of physical vulnerability.

When analyzing the hazard and risk index, mapped, respectively, in Fig. 4 and Fig. 5, it is interesting to see that when *PhyV* is included in the analysis, it considerably changes the results obtained for the hazard alone. This is very clear for the municipality of Setúbal, which stands out strikingly in the risk map in Fig. 5, but not in the hazard maps provided in Fig. 4. In fact, analyzing the proportion of the risk index that is explained by the hazard score, it represents less than ¼ of the final risk index, while exposure (in brown) and vulnerability (in violet) represent the vast proportion of the risk components. It should be highlighted that *E* and *PhyV* were only accounted for within the flood-susceptible areas. In the opposite case, the municipality of Loures, previously assigned a very high hazard Index, drops to an intermediate position in the risk index. This is explained by the proportionally low contribution of exposure and physical vulnerability of buildings.

By taking into account the data from the BGE and the BGRI, crossed with a precise assessment of susceptibility based on geomorphological criteria, the analysis carried out makes it possible to locate exactly where the buildings susceptible to flooding are, to know their main physical characteristics, and to estimate their resident population. For future works, these results can be aggregated at different levels of representation, such as statistical section, parish, or municipality, and thus provide an assessment with a view to defining inter-municipal risk management policies. Such policies should not be restricted to the sole focus on emergency and civil protection, as cross-sectoral and comprehensive responses to disaster risk reduction include domains such as social, environmental, urban planning, mobility, and land use management policies.

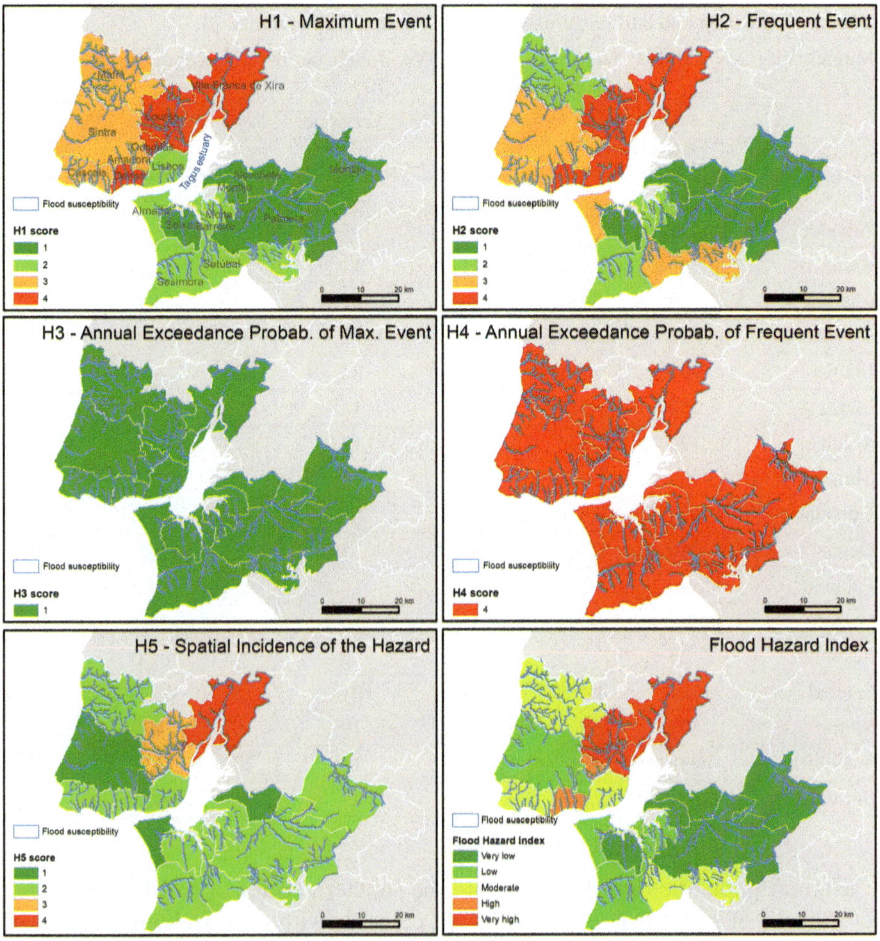

Fig. 4 Flood hazard parameters and index in the Lisbon Metropolitan area at the municipal level

4.2 Civil Parish Level

At the civil parish, 16 of the 118 territorial units don't have flood susceptibility significant enough to be considered. They are visible in Figs. 6a and 7a, represented with scores of 0 in flood hazard. In those 18 civil parishes, exposure and vulnerability (whether physical vulnerability of buildings or that of the resident population) are absent as well. Tagus river civil parishes, as well as those drained by the small watersheds north of Lisbon, present the highest hazard scores. Exposure of buildings is also high north of Lisbon (civil parishes of the Odivelas and Loures municipalities) and in the civil parishes at the mouth of small watercourses of Setúbal and Cascais municipalities (Fig. 6b; the name of municipalities can be checked in Figs. 1 and 5). Population exposure partly offsets that of buildings.

Table 2 Population and buildings' exposure and physical vulnerability

Municipality	No. Bldg	No. Inhab	Mean PV	Inhab. per building condition (parameter P7)			
				Very poor	Poor	Good	Very good
Alcochete	12	24	0.38	0	3	11	10
Almada	194	888	0.33	0	9	146	733
Amadora	28	226	0.23	7	0	29	190
Barreiro	46	190	0.41	11	6	62	111
Cascais	779	3304	0.30	67	46	1028	2163
Lisbon	1813	17,675	0.50	1062	922	6050	9641
Loures	991	3503	0.31	88	123	832	2460
Mafra	255	523	0.33	14	4	145	360
Moita	124	235	0.30	14	3	90	128
Montijo	39	66	0.35	2	2	30	32
Odivelas	1314	13,419	0.40	88	281	5508	7542
Oeiras	340	2726	0.39	25	82	589	2030
Palmela	336	876	0.31	7	4	128	737
Seixal	532	4691	0.37	15	77	941	3658
Sesimbra	5	4	0.25	0	0	0	4
Setúbal	1781	14,485	0.44	191	754	4623	8917
Sintra	577	5311	0.36	19	107	1479	3706
V. F. Xira	1035	4515	0.38	17	203	1335	2960
Sum/*Mean*	10,201	72,661	0.35	1627	2626	23,026	45,382

Highest population's vulnerability in the southern sector of LMA—fortunately, for the sake of a low-risk index—does not coincide with areas of high hazard (civil parishes in Almada, Seixal, or Barreiro municipalities), Fig. 7a and c.

Even with hazard being the only common component between flood risk at the civil parish using exposure and vulnerability of population (Fig. 8a) and exposure and vulnerability of buildings (Fig. 8b), the differences between the two maps are not as expressive as eventually expected. This fact can be explained by two arguments:

(a) Civil parishes where the vulnerability of people is high coincide with civil parishes where the physical vulnerability of buildings is equally high. In simple terms, this would mean that vulnerable persons tend to reside in vulnerable buildings and vice-versa;
(b) The role of vulnerability and of exposure, combined, is not strong enough to have an impact on the final risk score after their multiplication by hazard. We recall that equal weights are applied to all three components of risk.

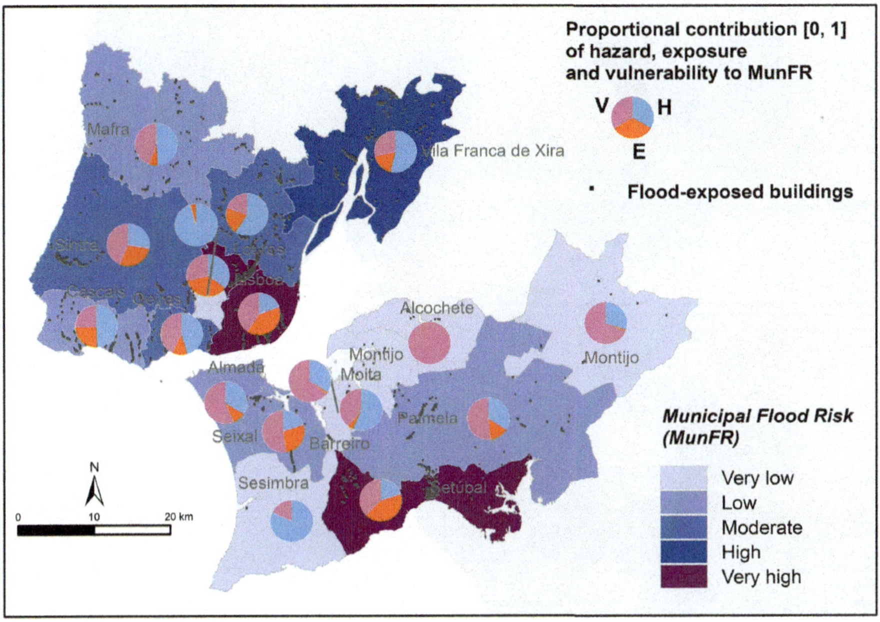

Fig. 5 Flood risk index as an expression of flood susceptibility, buildings' and population's exposure, and buildings' physical vulnerability, at the municipal level in the LMA

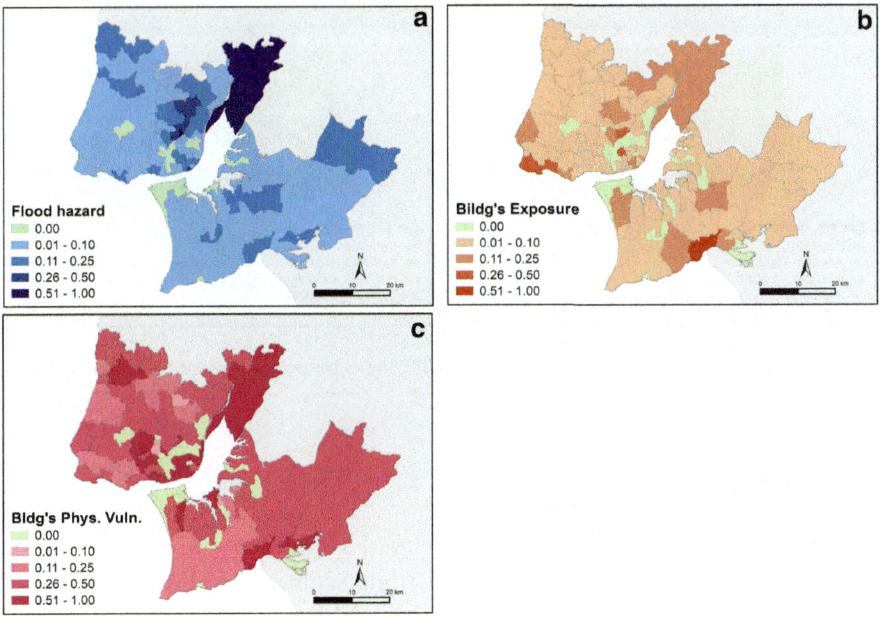

Fig. 6 Risk components at the civil parish level considering the exposure of buildings, in the LMA

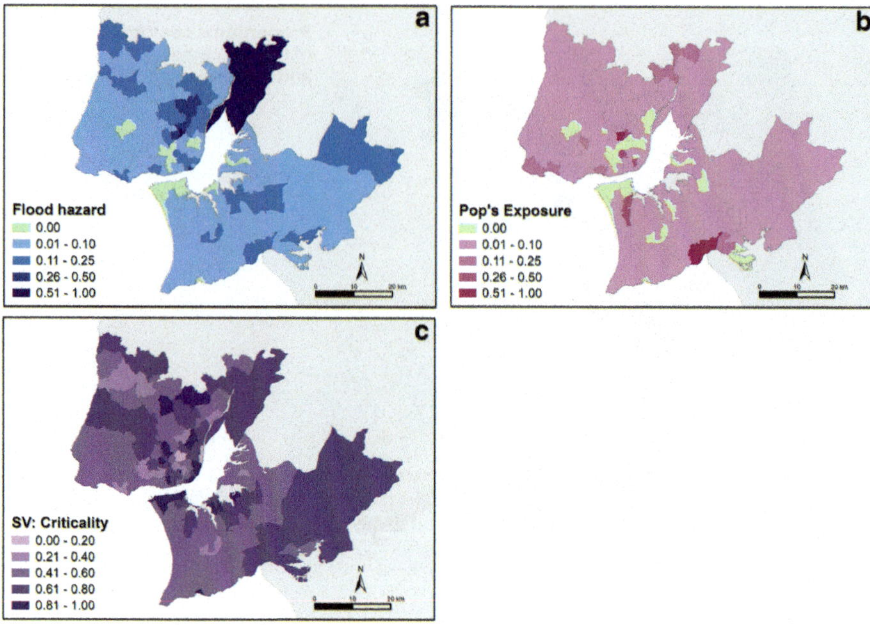

Fig. 7 Risk components at the civil parish level considering the exposure of population and its criticality in the LMA

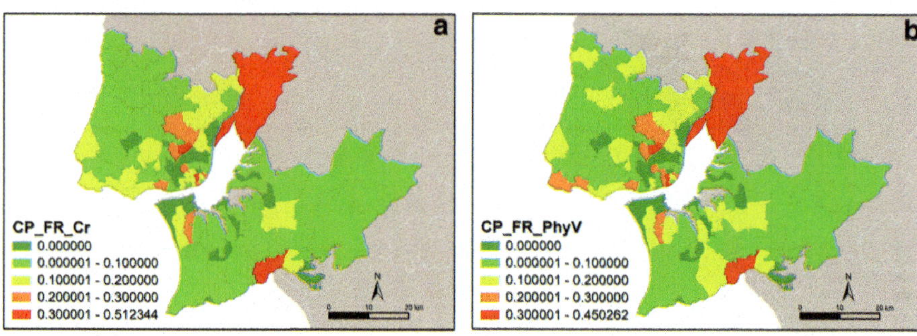

Fig. 8 Flood risk at the civil parish level considering the exposure of population (**a**) and buildings (**b**), in the LMA

4.3 Building Level

The LMA ASPFR areas differ significantly: Torres Vedras area, in the Mafra municipality, sums only five residential buildings, while Setúbal's ASPFR comprises 1411 (Table 3). The BPVF mean in the four EU's Flood Directive areas is 37.7, a value higher than the mean in the entire LMA (28.5), in a range from 0 to 100.

Table 3 Summary of results from the building-level cluster analysis in the ASPFR. The building's physical vulnerability to flooding (BPVF), flood depth and velocity, and vulnerability parameters (P1–P7) are expressed by the average within each cluster

ASPFR	Cluster	No. inhabitants	No. buildings	BPVF	Flood depth (m)	Veloc (m^2/s)	P1 (Age)	P2 (Floors)	P3 (Claddin mat)	P4 (Struc. sys.)	P6 (Expos.)	P7 (Repair Cond.)
Loures and Odivelas (L&O)	1	*2*	*4.0*	*75.3*	*0.4*	*0.1*	*92.5*	*77.5*	*100.0*	*100.0*	*32.5*	*77.5*
	2	*2866*	*204.0*	*51.2*	*0.8*	*0.3*	*49.7*	*77.1*	*48.1*	*41.9*	*15.7*	*33.5*
	3	*2003*	*559.0*	*26.1*	*0.4*	*0.2*	*37.8*	*24.0*	*10.9*	*20.6*	*28.2*	*26.2*
	4	*6562*	*550.0*	*38.6*	*0.4*	*0.2*	*39.3*	*72.8*	*12.1*	*21.3*	*28.8*	*23.4*
	ASPFR L&O	*11,432*	*1317.0*	*35.4*	*0.5*	*0.2*	*40.5*	*52.8*	*17.4*	*24.4*	*26.5*	*26.3*
Setúbal	1	*2*	*4.0*	*75.8*	*0.2*	*0.2*	*100.0*	*85.0*	*92.5*	*100.0*	*32.5*	*100.0*
	2	*2214*	*361.0*	*50.5*	*0.3*	*0.3*	*85.1*	*63.6*	*32.9*	*60.6*	*14.5*	*43.8*
	3	*794*	*342.0*	*28.3*	*0.1*	*0.2*	*56.1*	*19.7*	*10.0*	*29.3*	*27.9*	*31.6*
	4	*8283*	*704.0*	*39.9*	*0.2*	*0.3*	*56.3*	*57.9*	*13.8*	*39.2*	*20.1*	*30.7*
	ASPFR Set	*11,293*	*1411.0*	*39.9*	*0.2*	*0.3*	*63.7*	*50.2*	*18.0*	*42.4*	*20.6*	*34.5*
Tagus floodplain	3	*9*	*4.0*	*29.5*	*0.8*	*0.4*	*32.5*	*10.0*	*10.0*	*32.5*	*55.0*	*17.5*
	4	*30*	*29.0*	*34.7*	*1.5*	*0.6*	*39.0*	*12.1*	*11.0*	*40.0*	*100.0*	*16.2*
	ASPFR Tagus	*39*	*33.0*	*34.0*	*1.4*	*0.6*	*38.2*	*11.8*	*10.9*	*39.1*	*94.5*	*16.4*
Torres Vedras (T.V.)	3	*2*	*1.0*	*32.0*	*0.0*	*0.0*	*70.0*	*10.0*	*10.0*	*10.0*	*100.0*	*10.0*
	4	*9*	*4.0*	*34.0*	*0.2*	*0.1*	*10.0*	*40.0*	*10.0*	*10.0*	*100.0*	*10.0*
	ASPFR T.V.	*11*	*5.0*	*33.6*	*0.1*	*0.1*	*22.0*	*34.0*	*10.0*	*10.0*	*100.0*	*10.0*
LMA's ASPFR		*22,776*	*2766.0*	*37.7*	*0.3*	*0.2*	*52.3*	*50.9*	*17.6*	*33.8*	*24.4*	*30.3*

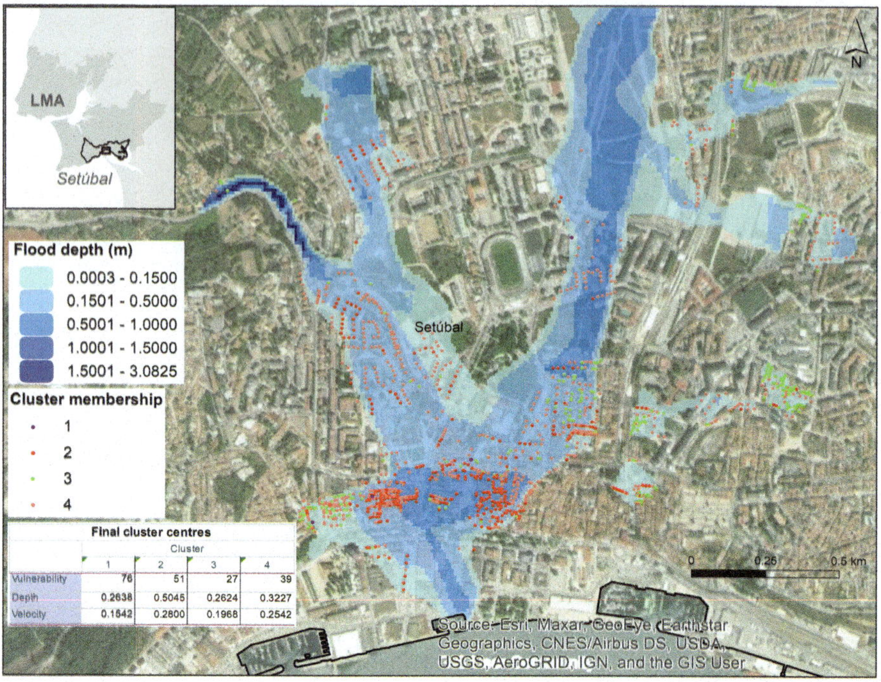

Fig. 9 Flood depths from the European Union's Floods directive and cluster membership of buildings in the city of Setúbal, Portugal

The resulting four clusters can be interpreted as follows: cluster 1 aggregates 8 buildings, characterized by high scores of physical vulnerability, while flood depth and velocity are, in general, low; cluster 2 aggregates 565 buildings, characterized by moderate scores of BPVF but generally high flood depths and velocities (the high hazard is the distinctive factor); cluster 3 (906 buildings) is the "safest" cluster, expressing low physical vulnerability, flood depths, and velocities; finally, cluster 4 (1287 buildings) represents an intermediate context between clusters 2 and 3. Figure 9 illustrates this in Setúbal city.

5 Final Remarks

Flood processes are complex in their triggering and conditioning factors. When societal aspects like those that define exposure and vulnerability are added to risk analysis, such complexity becomes even higher. Their representation at distinct scales, as performed in this study, denotes how specific hotspots of flood risk can remain hidden if detailed levels of analysis are neglected.

On the component of exposure and vulnerability alone, this analysis shows the potential of Census data in supporting local-level flood studies. It is fundamental to dispose of updated sociodemographic data at the statistical block and the building level, without the need to conduct dasymetric estimations from polygonal to point-type representations of residents. However, fieldwork remains essential, namely in verifying to which extent the buildings' characteristics are inadequate to the levels of flood hazard they are exposed to. Geomorphological analysis, along with hydraulic modeling, is key in preventing the generation of risk hotspots through both upstream and valley-level spatial planning, urban planning, and retrofitting.

In general, the results achieved allow concluding that the current Flood Directive's critical areas were sustainably selected. Three out of the four areas of significant potential flood risk present high scores of hazard, exposure, and vulnerability.

Risk practitioners and decision-makers at the metropolitan, municipal, and neighborhood levels have different information needs. Recent flood events (December 2021 in the region of Lisbon) highlighted the unpredictability of the exact location where rainfall intensities will record higher. Early warning systems are undoubtedly useful tools for emergency services. However, attention must be given to the flood risk components more accessible to societal control. They include the care with the artificialized stream courses in urban areas, the strict enforcement of land use restrictions in channels and valleys' bottoms and construction codes, and the care with the more vulnerable persons. Evidence shows that ground floors in specific locations should be prohibitive to the residential function. In summary, flood loss reduction implies the involvement of distinct actors acting at distinct scales and sectors related to risk governance, both upon the scale of the basin and the scale of the floodplain.

Acknowledgements The project 'MIT-RSC—Multi-risk Interactions Towards Resilient and Sustainable Cities' (MIT-EXPL/CS/0018/2019) leading to this work is co-financed by the ERDF—European Regional Development Fund through the Operational Program for Competitiveness and Internationalisation—COMPETE 2020, the North Portugal Regional Operational Program—NORTE 2020 and by the Portuguese Foundation for Science and Technology—GCT under the MIT Portugal Program at the 2019 PT call for Exploratory Proposals in 'Sustainable Cities'. RISKCOAST is funded by the Interreg Sudoe Programme (SOE3/P4/E0868). Pedro Pinto Santos is financed through FCT I.P., under the contract CEECIND/00268/2017.

References

1. Agliata R, Bortone A, Mollo L (2021) Indicator-based approach for the assessment of in-trinsic physical vulnerability of the built environment to hydro-meteorological hazards: Re-view of indicators and example of parameters selection for a sample area. Int J Disaster Risk Reduct 58:102199
2. Alfieri L, Dottori F, Betts R, Salamon P, Feyen, L (2018) Multi-model projections of river flood risk in Europe under global warming. Climate 6(1). https://doi.org/10.3390/cli6010006
3. APA (2015) Flood directive's risk assessment. Available at https://sniamb.apambiente.pt/content/geo-visualizador

4. Aven T, Renn O (2010) Risk management and governance: Concepts, guidelines and applications. Springer. https://doi.org/10.1007/978-3-642-13926-0
5. Choi HI (2019) Assessment of Aggregation Frameworks for Composite Indicators in Measuring Flood Vulnerability to Climate Change. Sci Rep 9(1):1–14. https://doi.org/10.1038/s41598-019-55994-y
6. EEA (2017) Climate change, impacts and vulnerability in Europe 2016. EEA Report No 1/2017. European Environment Agency, p 419
7. Fell R, Corominas J, Bonnard C, Cascini L, Leroi E, Savage WZ (2008) Guidelines for landslide susceptibility, hazard and risk zoning for land-use planning. Eng Geol 102:99–111
8. Garcia RAC, Oliveira SC, Zêzere JL (2015) Assessing population exposure for landslide risk analysis using dasymetric cartography. Nat Hazard 12:2769–2782
9. IPCC—Working Group II (2022) Climate change 2022—6th assessment report. Impacts, adaptation and vulnerability. Summary for policy makers. Working group II. Intergovernmental panel on climate change
10. Kappes MS, Papathoma-Köhle M, Keiler M (2012) Assessing physical vulnerability for multi-hazards using an indicator-based methodology. Appl Geogr 32(2):577–590
11. Leal M, Reis E, Pereira S, Santos PP (2021) Physical vulnerability assessment to flash floods using an indicator-based methodology based on building properties and flow parameters. J Flood Risk Manag:1–19. https://doi.org/10.1111/jfr3.12712
12. Lindersson S, Brandimarte L, Märd J, Di Baldassarre G (2021) Global riverine flood risk - How do hydrogeomorphic floodplain maps compare to flood hazard maps? Nat Hazard 21(10):2921–2948. https://doi.org/10.5194/nhess-21-2921-2021
13. Loats R, Petrascheck A (1997) Dangers naturels. Recommandations 1997. Prise en compte des dangers dus aux crues dans le cadre des activités de l'aménagement du territoire. Office fédéral de l'économie des eaux (OFEE) Office fédéral de l'aménagement du territoire (OFAT) Office fédéral de l'environnement, des forêts et du paysage (OFEFP). Bern, p 32
14. Marin-Ferrer M, Vernaccini L, Poljansek K (2017) Index for risk management inform concept and methodology report—Version 2017, EUR 28655 EN. https://doi.org/10.2760/094023
15. Mazzoleni M, Mård J, Rusca M, Odongo V, Lindersson S, Di Baldassarre G (2020). Flood-plains in the Anthropocene: A global analysis of the interplay between human population, built environment and flood severity. Water Resour Res 57(2):e2020WR027744. https://doi.org/10.1029/2020WR027744
16. Mendes JM, Tavares AO, Cunha L, Freiria S (2011) A vulnerabilidade social aos perigos naturais e tecnológicos em Portugal [Social vulnerability to natural and technological hazards in Portugal]. Revista Crítica de Ciências Sociais 93:95–128
17. Mendes JM, Tavares AO, Santos PP (2019) Social vulnerability and local level assessments: a new approach for planning. International Journal of Disaster Resilience in the Built Environment 11(1):15–43. https://doi.org/10.1108/IJDRBE-10-2019-0069
18. Miranda FN, Ferreira TM (2020) A simplified approach for flood vulnerability assessment of historic sites. Nat Hazards 96(2):713–730. https://doi.org/10.1007/s11069-018-03565-1
19. Moel H, van Alphen J, Aerts JCJH (2009) Flood maps in Europe - methods, availability and use. Nat Hazards Earth Syst Sci 9(2):289–301
20. Moel H, Jongman B, Kreibich H, Merz B, Penning.Rowsell E, Ward PJ (2015) Flood risk assessments at different spatial scales. Mitig Adapt Strateg Glob Change 20:865–890. https://doi.org/10.1007/s11027-015-9654-z
21. NATO (2006) Flood risk management: Hazards, vulnerability and mitigation measures. NATO Advanced research workshop on flood risk management. In: Schanze J, Zemen E, Marsalek J (eds). NATO Science Series, vol. 67. p 320
22. O'Keefe P, Westgate K, Wisner B (1976) Taking naturalness out of natural disasters. Nature 260:566–567
23. Pereira S, Santos PP, Zêzere JL, Tavares AO, Garcia RAC, Oliveira SC (2020) A landslide risk index for municipal land use planning in Portugal. Sci Total Environ 735:139463. https://doi.org/10.1016/j.scitotenv.2020.139463

24. Rougier J, Sparks S, Hill L (2013) Risk and uncertainty assessment for natural hazards. Cambridge University Press, Cambridge
25. Saber M, Abdrabo KI, Habiba OM, Kantosh SA, Sumi T (2020) Impacts of triple factors on flash flood vulnerability in Egypt: Urban growth, extreme climate, and mismanagement. Geosciences 10(1):24. https://doi.org/10.3390/geosciences10010024
26. Santos PP, Pereira S, Rocha J, Reis E, Santos M, Oliveira SC, Garcia R, Melo R (2022) Zezere JL (2022) The role of susceptibility, exposure and vulnerability as drivers of flood disaster risk at the parish level. Environ Earth Sci 81:465. https://doi.org/10.1007/s12665-022-10589-1
27. Silva M, Pereira S (2014) Assessment of physical vulnerability and potential losses of buildings due to shallow slides. Nat Hazards 72:1029–1050. https://doi.org/10.1007/s11069-014-1052-4
28. Stephenson V, D'Ayala D (2014) A new approach to flood vulnerability assessment for historic buildings in England. Nat Hazard 14(5):1035–1048. https://doi.org/10.5194/nhess-14-1035-2014
29. UNDRR (2019) Global Assessment Report 2019. UN Office for disaster risk reduction. Geneva, Switzerland
30. WBGU (2000) World in transition. Strategies for managing global environmental risks—Annual Report 1998. German Advisory Council on global change. Springer, p 366
31. Wing OEJ, Bates PD, Smith AM, Sampson CC, Johnson KA, Fargione J, Morefield P (2018) Estimates of present and future flood risk in the conterminous United States. Environ Res Lett 13(3). https://doi.org/10.1088/1748-9326/aaac65
32. Wisner B, Blaikie P, Cannon T, Davis, I (2004) At Risk. Natural Hazards, people's vulnerability and disasters, 2nd edn. Routledge, London
33. Zahmatkesh Z, Karamouz M (2017) An uncertainty-based framework to quantifying climate change impacts on coastal flood vulnerability: case study of New York City. Environ Monit Assess 189(11). https://doi.org/10.1007/s10661-017-6282-y
34. Zêzere JL, Pereira S, Tavares AO, Bateira C, Trigo RM, Quaresma I, Santos PP, Santos M, Verde J (2014) DISASTER: A GIS database on hydro-geomorphologic disasters in Portugal. Nat Hazards 72(2):503–532

Chapter 5
Seismic Vulnerability and Risk Assessment of the Lisbon Metropolitan Area

Maria Xofi, José Carlos Domingues, and Paulo B. Lourenço

Abstract This chapter presents an index-based seismic vulnerability assessment methodology, based on the GNDT level II approach, to be used in the seismic risk assessment of the residential building stock in the Lisbon Metropolitan Area. The buildings' physical vulnerability and exposure were defined using data from the 2011 national population and housing Census. The seismic vulnerability assessment methodology was then applied to the 292,978 reinforced concrete and 152,916 unreinforced masonry buildings within the Lisbon Metropolitan Area. A GIS tool was afterwards used to combine the vulnerability results and the seismic hazard data, allowing identifying and analyzing the areas of higher vulnerability and risk, as a way to improve the ability of stakeholders to devise efficient resilience improvement and disaster mitigation plans.

Keywords Seismic vulnerability · Vulnerability assessment · Seismic risk · Urban scale · Metropolitan scale

1 Introduction

Over the past few decades, damages caused by natural hazards on the building stock and infrastructure systems with associated economic and human losses have gained considerable attention within the research community and among stakeholders and risk mitigation managers. Among the several natural hazards that can impact human societies, such as earthquakes, floods, landslides, and fires, earthquakes have been the most catastrophic phenomena in terms of economic losses and casualties.

M. Xofi
Faculty of Engineering Technology, University of Twente, Enschede, The Netherlands
e-mail: m.xofi@utwente.nl

M. Xofi · J. C. Domingues (✉) · P. B. Lourenço
Institute for Sustainability and Innovation in Structural Engineering (ISISE), Department of Civil Engineering, University of Minho, Guimarães, Portugal
e-mail: jcarl.domingues@gmail.com

T. M. Ferreira (ed.), *Multi-risk Interactions Towards Resilient and Sustainable Cities*, Advances in Sustainability Science and Technology,
https://doi.org/10.1007/978-981-99-0745-8_5

Based on the CATDAT damaging earthquake database, the 2010 Haiti earthquake, for instance, resulted in a death toll estimated to be in the range of 46,000–316,000 casualties. The Tohoku earthquake in 2011 in Japan caused 20,475 fatalities and left 1.108 million people homeless, while the economic losses amounted to $140 billion. Yet on the economic consequences, the financial loss caused in Turkey after the occurrence of the Van earthquake in 2011 was $2.2 billion, and about $1.7 billion when the Sikkim earthquake struck India in 2011 [10]. All of these are just a few examples of extreme and catastrophic events caused by single and multi-hazard events, such as the Tohoku earthquake, which was also combined with a tsunami.

In this chapter, a seismic vulnerability assessment methodology based on an index-based approach is developed and validated for the Lisbon Metropolitan Area (LMA), using data from the 2011 national Census survey. This information is used not only to categorize the existing building stock but also to define the parameters to be considered in the vulnerability analysis. The formulation developed in this research is based on the GNDT level II approach, which combines a typological approach and a vulnerability index-based estimation and is based on data from post-earthquake damage observations and information concerning the typological characterization of the building stock, subsequently translated into a few empirical parameters (GNDT, 1994). The seismic vulnerability results and the seismic hazard components were then integrated into a Geographic Information System (GIS) tool developed in the open-source software QGIS to obtain the different seismic risk levels for the municipalities of LMA.

2 Conceptual Framework of the Risk Assessment Approach

As mentioned before, this chapter focuses on the seismic vulnerability of the residential building stock of the Lisbon Metropolitan Area, which comprises both reinforced concrete (RC) and unreinforced masonry (URM) buildings. In particular, this study aims to assess the seismic vulnerability parameters and present a simplified methodology for evaluating the seismic vulnerability at the urban scale. Based on the seismic vulnerability results discussed in the present chapter and available seismic hazard maps for the area, a seismic risk assessment of the building stock in LMA is carried out. Three core components were considered: seismic hazard, exposure (which is related to the presence of people, buildings, and infrastructures that could be affected) and the seismic vulnerability of buildings. The methodological framework adopted here for the risk assessment is based on the same approach carried out by Ferreira and Santos [7].

The level of hazard is obtained by combining the earthquake intensity, the Peak Ground Acceleration (PGA), and the soil classes for the LMA, as will be described in Sect. 2.1. A seismic vulnerability assessment methodology is afterwards presented in

Table 1 Seismic risk matrix

Vulnerability	Hazard			
	Low	Moderate	High	Very high
Low	Low	Low	Moderate	High
Moderate	Low	Moderate	High	Very high
High	Moderate	High	Very high	Extreme
Very high	High	Very high	Extreme	Extreme

Sect. 2.3. Finally, a seismic risk assessment is then conducted using a vulnerability-hazard matrix that relates each building's vulnerability with the level of the exposed hazard, as shown in Table 1.

2.1 Characterization of the Seismic Hazard

Southern Portugal and, in particular, the Lisbon Metropolitan Area, the region of Algarve and the Azores Islands, are considered the three most critical seismic regions in Portugal. The seismicity of these regions is associated with the fact that they are located near the Eurasia-Africa plate boundary facing the Atlantic Ocean, which makes all these regions prone to high-devastating offshore and onshore earthquakes [2, 18]. Several large earthquakes have occurred, including the 1755 event and an M7.6 and an M7.9 earthquake in 1816 and 1969, respectively.

The 1755 Lisbon earthquake, with an estimated magnitude of 8.5–9, was the largest known historic earthquake to impact Europe and northern Africa. Various studies have located the epicenter at about 300–400 km southwest of Lisbon in the Gorringe Bank (GB), along the Africa-Eurasia plate boundary. According to Franco and Shen-Tu [8], this plate movement has caused various types of deformation: extension along the Terceira Ridge around the Azores Islands in the west, strike-slip along the Gloria Fault in the central segment; and convergence and collision in the Gorringe Bank and the Strait of Gibraltar to the east. The collision stress not only loads up major structures within the collision zone, where great earthquakes have been originated, including the 1755 Lisbon earthquake but also triggers many onshore crustal faults at the Iberia and Africa continent margins. Several onshore faults in southwestern Portugal are seismically active. Most notable is the active fault zone in the Lower Tagus Valley (LTV), which has gained considerable attention in recent years [17] as it passes through Lisbon, Portugal's capital and most populous city. Several large historical earthquakes, including the 1531 M7, the 1909 M6.3, and possibly the 1344 $M > 6$, occurred on the LTV, causing significant damage in the Lisbon area.

Franco and Shen-Tu [8] proposed that the offshore 1755 Lisbon earthquake triggered an onshore rupture on the LTV fault. Therefore, the return period of magnitude VI–VII earthquakes along the LTV could be as short as 150–200 years, making

Lisbon the highest-risk area in Portugal. Hence, from a risk assessment perspective, an earthquake of magnitude VI–VII close to Lisbon could cause comparable or even greater damage than a stronger shock offshore.

The seismic hazard in the Lisbon and Vale do Tejo Region is typically high due not only to the proximity of active faults of the mainland to the southwest and south, which have the potential to generate strong regional earthquakes but also to the fault (or fault zone) of the lower Tagus valley. The 1986 Portuguese code for actions establishes zoning of seismic risk in mainland Portugal and defines the safety factors to be applied in the construction of buildings and bridges in each of the four zones identified. The Lisbon and Vale do Tejo (Tagus Valley) Region are included in zones A and B of the referred zoning, which corresponds to the areas of greatest risk. More recently, Eurocode 8 has updated the general rules and seismic actions for buildings in the framework of earthquake-resistant structural design, with the Lisbon and Vale do Tejo Region included in zones 1.3, 1.4, and 1.5 for Type 1 seismic actions (far-field earthquakes), and zones 2.3 and 2.4 for Type 2 seismic actions (near field earthquake).

The seismic hazard in the territory of Lisbon and Tagus Valley was defined by crossing the maximum seismic intensities (obtained from the seismic intensity map created by the Portuguese Institute for Sea and Atmosphere) with the distribution of maximum accelerations (PGA—Peak Ground Acceleration) for a return period of 475 years (obtained from [13]. Additionally, the soil effects capable of producing an amplification of the seismic actions were defined from the distribution of non-consolidated sedimentary geological formations, represented in the Geological Map of Portugal (from the National Laboratory of Energy and Geology) and also from the zoning strips of 100 m around active faults (extracted from the Neotectonic Map of Portugal). Figures 1, 2, and 3 illustrate the LMA territory, representing maximum seismic intensity, peak ground acceleration, and quaternary deposits.

Therefore, and taking into account the seismic intensity maps, it is estimated that about 40% of the LVT territory integrates the class corresponding to the maximum intensity level VIII (Modified Mercalli Scale, 1956), which represents a scenario of ruin, while the remaining 60% of the territory belong to classes corresponding to intensity degrees IX (disaster) and X (collapse). Regarding the distribution of maximum ground accelerations, there is a PGA of 3.2–4.0 m/s^2 and 2.4–3.2 m/s^2 in 41.4 and 46.3% of the LVT Region, respectively.

The assessment of exposure (i.e., buildings and residents) to seismic risk was carried out considering four levels of seismic hazard (low, moderate, high, and very high), having as a reference the reality of the entire Portuguese territory. The low hazard class corresponds to the crossing of grade VIII seismic intensity zones with PGA of 2.4–3.2 m/s^2 and 1.6–2.4 m/s^2, respectively. There is no seismic amplification induced by poorly consolidated sedimentary deposits or active faults in this hazard class, which is present in 23.5% of the territory.

The moderate hazard class corresponds to the crossing of the following zones of seismic intensity, maximum accelerations, and seismic amplification: (i) seismic intensities of grade IX and X with PGA from 2.4 to 3.2 m/s^2; (ii) grade VIII and PGA from 3.2 to 4.0 m/s^2; (iii) seismic intensities grade VIII seismic intensity and PGA from 2.4 to 3.2 m/s^2 or 1.6 to 2.4 m/s^2 with the presence of poorly consolidated

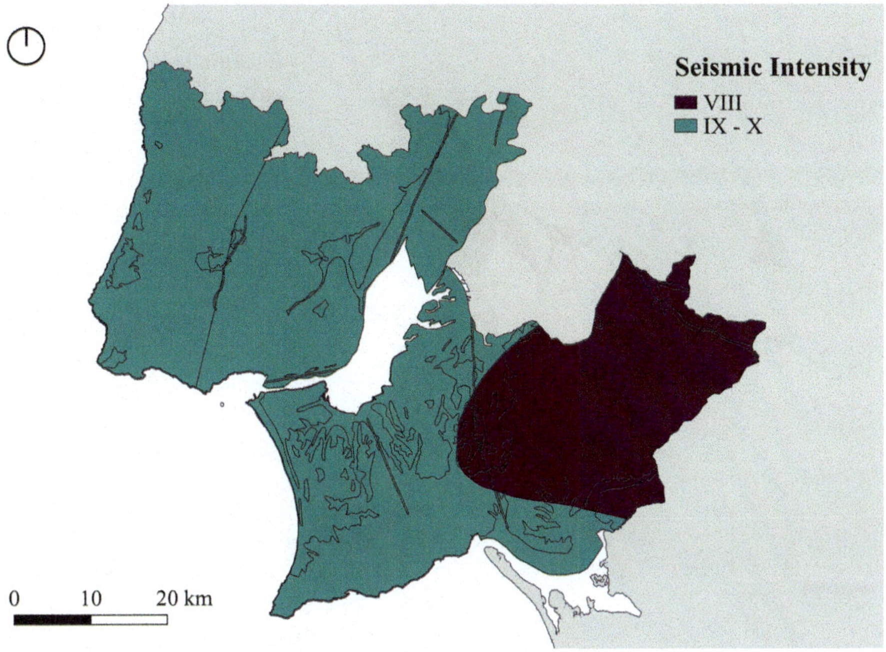

Fig. 1 Maximum seismic intensity in LMA

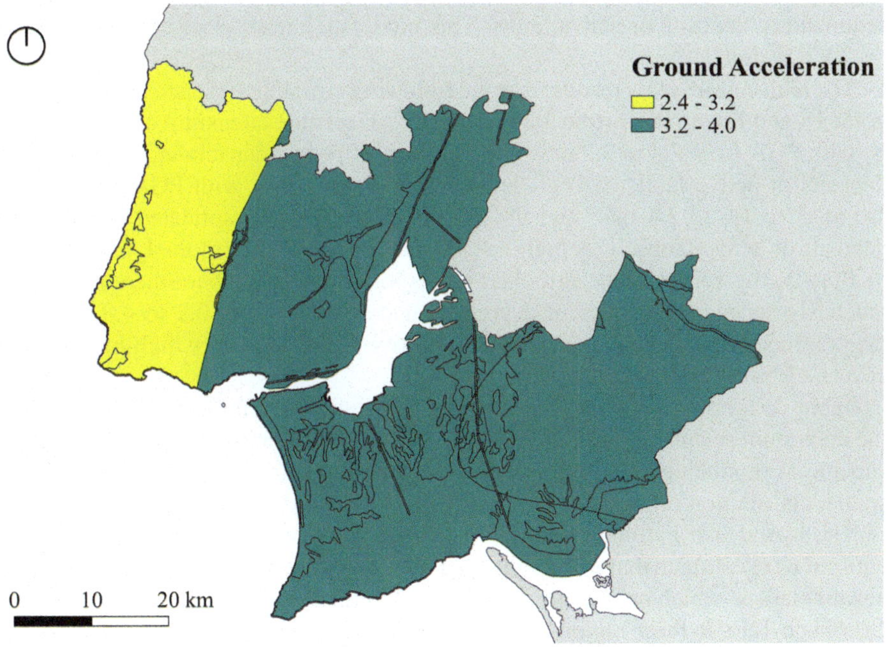

Fig. 2 Maximum peak ground acceleration in LMA, for 475 years

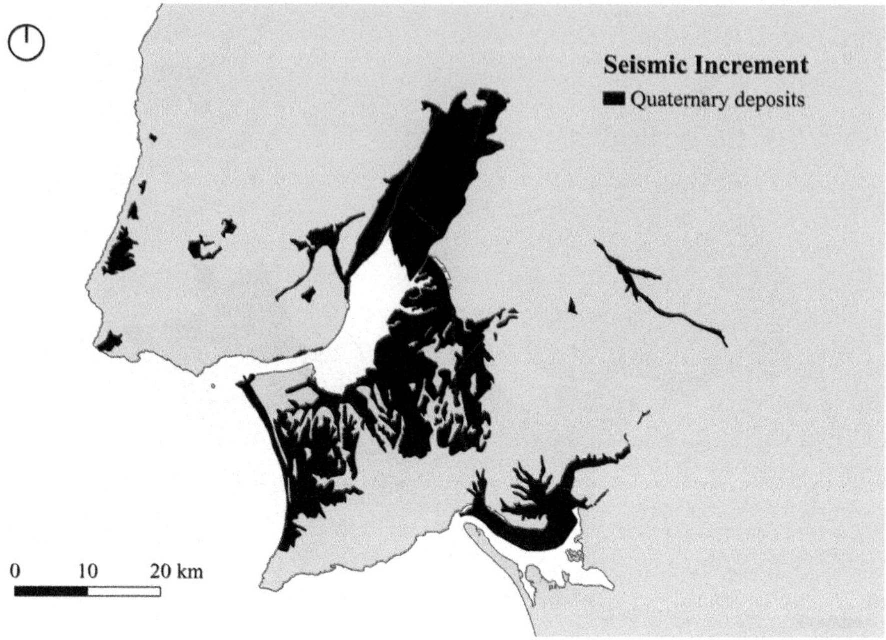

Fig. 3 Quaternary deposits within LMA

sedimentary deposits or active faults. This moderate hazard class covers 40.5% of the LVT region.

The high-hazard class results from the following crossing: (i) seismic intensities of grade IX and X with PGA from 3.2 to 4.0 m/s^2; (ii) seismic intensities of grade IX and X, with PGA from 2.4 to 3.2 m/s^2 and presence of poorly consolidated sedimentary deposits or active faults; (iii) grade VIII seismic intensity with PGA from 3.2 to 4.0 m/s^2 or 1.6 to 2.4 m/s^2 and the presence of poorly consolidated sedimentary deposits or active faults. This high-hazard class covers 23.5% of the LVT region.

Finally, the very high-hazard class corresponds to areas where there are simultaneously seismic intensities of degree IX and X, PGA from 3.2 to 4.0 m/s^2 and the presence of unconsolidated deposits or active faults. This very high-hazard class covers 12.5% of the LVT region.

Table 2 summarizes the definition of the seismic hazard classes, summarizing the maximum seismic intensities, the Peak Ground Acceleration, and the soil effects capable of producing an amplification of the seismic actions. Figure 4 presents the spatial distribution of the hazard classes in the Lisbon Metropolitan Area.

The spatial distribution of maximum intensity, PGA, and seismic increment zones allowed us to estimate that 36% of the LVT Region integrates the high and very high-hazard classes, which are mainly concentrated in the Lisbon Metropolitan Area and Lezíria do Tejo. In these regions, not only the influence of seismicity is reflected, but also the fault (or fault zone) of the lower Tagus valley. Last but not least, it should be

Table 2 Seismic hazard classes

Intensity level (Modified mercalli scale)	Peak ground acceleration		
	1.6–2.4	2.4–3.2	3.2–4.0
Seismic amplification			
VIII	Moderate	Moderate	High
IX	–	High	Very high
X	–	High	Very high
No seismic amplification			
VIII	Low	Low	Moderate
IX	–	Moderate	High
X	–	Moderate	High

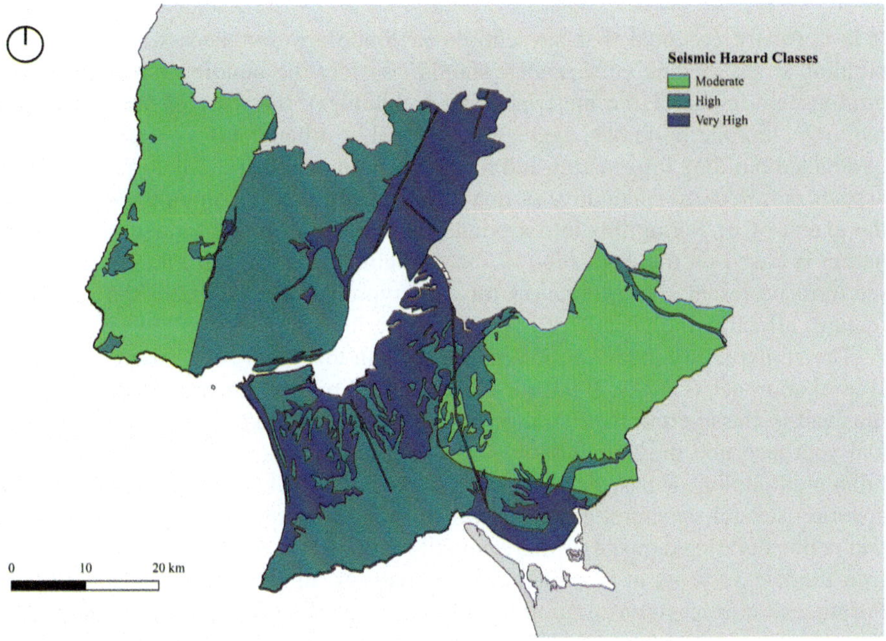

Fig. 4 Spatial distribution of the seismic hazard classes within the LMA

mentioned that the very high seismic hazard class covers a total of 15 municipalities, such as Alcochete, Almada, Barreiro, Loures, Moita, Montijo (Afonsoeiro), Seixal, Sesimbra, Setubal, and Vila Franca de Xira, while the high seismic hazard class includes 22 municipalities, for instance, Amadora, Cascais, Lisbon, Mafra, Odivelas, Oeiras, Palmela, Sintra, and Montijo (Canha).

2.2 *Identification and Characterization of the Exposed Elements*

A detailed sociodemographic and typological characterization was conducted for the whole extent of the Lisbon Metropolitan Area in Chap. 3 of this book. Information from the 2011 national Census survey was used not only to describe and categorize the existing building stock but also to define the parameters to be considered in the vulnerability analysis.

2.3 *Seismic Vulnerability Assessment of the Lisbon Metropolitan Area's Building Stock*

It is normally accepted that any chosen methodology for the assessment of the building stock seismic vulnerability should consider the building typology (structural system, construction materials, and techniques), the scale of the assessment required (singular buildings, aggregate, regional or urban area), as well as the level of detailed building information and resources available [3, 12, 16]. Typically, these aspects do affect the qualitative or quantitative nature of the adopted approach and the extent of it. Taking this into account, the methodology developed and applied herein is based on the evaluation of the available Census 2011 building data with the most representative parameters for each typology that describes the structural features affecting the seismic response of the buildings.

The vulnerability index formulation applied in this work is based on the GNDT II level approach (GNDT, 1994) for the vulnerability assessment of masonry buildings and is classified in the literature as a hybrid approach combining the typological approach and the vulnerability index-based estimation. The seismic vulnerability methodology is based on post-earthquake damage observation and survey data covering several structural elements, which focuses on the most important aspects that define building damage. It was originally proposed in Italy and has been applied over the last 30 years in many large-scale analyses. In 2011, it was adapted to the Portuguese masonry construction by Vicente et al. [15] and improved further by introducing more detailed analysis for cases where adequate building data exist with new parameters related to the building's position and interaction between adjacent structures. The methodology has been enhanced by Ferreira et al. [5], who calibrated the method according to damage data collected following the 1998 Azores earthquake. It has been used in vulnerability assessment campaigns in different historical city centers, for instance, Coimbra [15], Seixal [4], Faro [11], and Leiria [1].

Similarly to the original proposal, the methodology developed in this work can be used to obtain a seismic vulnerability index based on the evaluation of a few parameters of empirical nature. Each parameter corresponds to a specific structural characteristic that affects the seismic response of a building with a corresponding vulnerability class that is most applicable. The adopted parameters are presented in

Table 3 Vulnerability parameters for RC buildings, with vulnerability classes and weights

Parameters	Class, C_{vi}				Weight p_i
	A	B	C	D	
P1. Building position	0	5	20	50	1.50
P2. Period of construction	–	5	20	50	2.00
P3. Number of stories	0	5	20	50	0.75
P4. Ground plan configuration	0	–	–	50	2.50

Table 4 Vulnerability parameters for URM buildings, with vulnerability classes and weights

Parameters	Class, C_{vi}				Weight p_i
	A	B	C	D	
P1. Structural system	–	5	20	50	2.50
P2. Period of construction	0	5	20	50	0.75
P3. Building position	0	5	20	50	1.00
P4. Number of stories	0	5	20	50	0.50
P5. Ground plan configuration	0	–	–	50	0.50

Tables 3 and 4 for the case of reinforced concrete (RC) and unreinforced masonry (URM) buildings, respectively, and further described in Sects. 2.3.1 and 2.3.2. These parameters are classified according to four vulnerability classes (C_{vi}) of A, B, C, and D and are associated with a weight, p_i, which defines the relative importance of each parameter to the overall seismic vulnerability of the building.

A total vulnerability index, I_v^*, is calculated using Eq. (1) by computing the weighted sum of the parameters multiplied by their specific weight assigned as a meaning of importance in terms of seismic response.

$$I_{\mathrm{v}f}^* = \sum_{i=1}^{n} C_{\mathrm{v}i}\, p_i \tag{1}$$

For easier interpretation and use, the vulnerability index, I_v^*, is then normalized to range between 0 and 100, assuming from that moment on the notation. The vulnerability level is given directly by the seismic vulnerability index. Vulnerability indexes ranging between 25 and 50 and between 50 and 75 were defined as "moderate" and "high", respectively. It should be noted that the value of 50 (the boundary value between "moderate" and "high") is often used in index-based vulnerability assessment approaches as a threshold for high vulnerability, which is applicable in the current study case.

2.3.1 Vulnerability Parameters for RC Buildings

From the analysis of Table 3, it is evident that parameters P2 (related to the period of construction), with a weight of 2.0, and P4 (regarding the ground plan configuration, i.e., the occurrence of possible soft-story mechanism at the ground floor level), with a weight of 2.5, are the ones that mainly influence the seismic vulnerability of the RC building stock. Nevertheless, and even though the weights of the other two parameters, P1 (building position) and P3 (number of stories), are the lowest ones, at 1.5 and 0.75, respectively; their contribution is still essential because together with the parameters P2 and P4 they affect the overall building seismic response. This subsection evaluates in more detail each of the four building parameters and the vulnerability classes, which are described in Table 5.

As previously mentioned, parameter P1 is related to the building's relative position and interaction with the adjacent neighboring buildings leading to possible pounding effects. According to the available Census data, four different building positions were considered, with each one corresponding to a different vulnerability class. The "Isolated" buildings (class B) represent almost 40%, while the "Classic" buildings (class C) account for 30% of the LMA building stock.

Table 5 Classes definition for the RC buildings

Parameters	
Parameter P1. Building position	
A	Terraced buildings
B	Isolated buildings
C	Classic buildings
D	Semi-detached or edge buildings
Parameter P2. Period of construction	
A	After 2018
B	1981–2018
C	1961–1980
D	Before 1960
Parameter P3. Number of stories	
A	1–2
B	3–7
C	8–15
D	> 15
Parameter P4. Ground plan configuration	
A	Regular buildings
B	–
C	–
D	Irregular buildings

Parameter P2 addresses the period of design and construction of the building relative to the seismic design code provisions. Given the absence of any field data related to the quality of RC building design and their structural systems, the vulnerability classes were ranked conservatively based on the historical evolution of the Portuguese seismic codes and regulations. Hence, class D was considered the most vulnerable for buildings without any seismic design built before the 1960s; Class C for buildings between 1961 and 1983, with minor seismic provisions; class B for buildings built between 1983 and 2011, based on current and modern seismic codes; and class A was assigned for buildings after 2018 (least vulnerable), which are not part of the available Census 2011 survey data. According to the abovementioned classifications, the distribution of the RC building stock within the Lisbon Metropolitan Area indicates 54% for buildings designed and built with the current seismic codes (Class B); 34% based on minor seismic provisions (Class C), and 12% for buildings with no seismic design code considerations (Class D).

Parameter P3 is referred to the number of stories (i.e., the height of the building), which affects the seismic vulnerability of the building in several ways. For instance, the taller the building, the higher becomes the top story drift ratio demands, which requires, in turn, appropriate seismic detailing to ensure sufficient displacement ductility capacity of the building. Therefore, the seismic vulnerability tends to increase with the additional number of stories. According to the distribution of buildings as per the classification of Parameter P3, the majority of RC buildings (63%) are low rise (1–2 floors), about 33% are mid-rise (3–7 floors), and 4% are high-rise buildings (8–15 floors).

Finally, parameter P4 evaluates the vertical irregularity when the layout of vertical structural elements (walls or columns) on the ground floor level changes about the upper floors creating an open plan area for commercial use or car parking spaces. This building typology, often known as "piloti" building, creates a possible soft-story mechanism at ground level, with a difference in strength and stiffness between the ground and the upper floors. As a result, this leads to a concentration of higher force demands to lesser structural elements and also to increased deformation demands at the first-floor level, which can lead to a brittle (shear) type of failure (i.e., unexpected collapse) mechanism on the ground level columns or walls. Consequently, class D is assigned as the most vulnerable given this particular building typology with a potential soft-storey mechanism at the ground floor level. The non-evaluated portion of the building stock has also been classified conservatively in class D, considering the significant number of buildings included in the sample (70%) and the uncertainty of what they may represent. On the other hand, Class A represents the least vulnerable cases, regular buildings, which account for 9% of the buildings.

2.3.2 Vulnerability Parameters for URM Buildings

The weights assigned to parameters for the vulnerability assessment of unreinforced masonry buildings that are presented in Table 4 show that the parameters that influence the seismic vulnerability of the URM buildings the most are parameter P1, which refers to the structural systems (including construction materials) and parameter P3 regarding the building's position to adjacent buildings and within the building aggregate. The parameter weights corresponding to the above-mentioned are 2.5 and 1.0, respectively. Even though the weights (p_i) of the other three parameters of P2 (Period of Construction), P4 (Number of stories), and P5 (Ground plan configuration) are the lowest ones, with 0.75 on the former and 0.5 on the latter ones, their contribution is still essential because together with the other three, they can affect the overall seismic response of the building. This subsection evaluates in more detail each of the four building parameters and the vulnerability classes, which are described in Table 6.

Parameter P1 refers to the type of structural system classifying the URM in different typologies and consequently with varying levels of vulnerability. Considering the age of construction with the material decay and older construction techniques, adobe or rubble stone buildings were considered the most vulnerable and assigned to Class D. Then, the mixed masonry-RC buildings with the concrete floor

Table 6 Classes definition for the URM buildings

Parameter P1. Structural system	
A	–
B	URM building with timber floor
C	URM building with concrete floor slabs
D	Adobe or rubble stone building
Parameter P2. Period of construction	
A	1981–2011
B	1961–1980
C	1945–1960
D	Before 1945
Parameter P3. Building position	
A	Terraced building
B	Isolated building
C	Classic building
D	Semi-detached or edge building
Parameter P4. Number of stories	
A	1
B	2–3
C	4–5
D	>6
Parameter P5. Ground plan configuration	
A	Regular building
B	–
C	–
D	Irregular building

slab, commonly known as "Placa" buildings, are assigned in Class C vulnerability, given their inherent fragility and all insufficient structural alterations with poor reinforcing detailing for seismic actions. The least vulnerable Class B refers to the traditional URM buildings with timber flooring, which are typically characterized by flexible floor diaphragms with inadequate connections to masonry walls. From the distribution of buildings across the LMA, it appears that the majority of the buildings (about 64%) represent the "placa" buildings (Class C), with the traditional URM with timber floors accounting for 31% (in Class B), and the remaining portion of 5% being adobe or rubble masonry buildings.

Parameter P2 addresses the period of design and construction of the building relative to the introduction of the main seismic design codes. The vulnerability classes were ranked considering the main benchmarks of the Portuguese seismic codes and the aging of the construction materials in terms of material decay and possible lack of maintenance through time. Hence, Class D was considered the most vulnerable for buildings before 1945, and Class C for buildings built during 1945–1960, which both represent the pre-code period. The least vulnerable classes, A and B, correspond to the most recent building construction during the periods of 1981–2011 (post-code: > 1983) and 1961–1980 (mid-code: 1958–1983), respectively. Except for the buildings built during 1945–1960, which account for 18%, the rest of the building distribution ranges from 26 to 28%, showing an overall constant trend up until 2011.

Parameter P3 is related to the building's relative position and interaction with the adjacent buildings creating a possible pounding effect. According to the available Census data, four different building positions were considered, with each one corresponding to a different vulnerability class. For instance, terraced buildings are classified in vulnerability class A, isolated buildings in vulnerability class B, the "classic" buildings are classified with vulnerability Class C and the semi-detached or edge buildings are assigned to Class D. The distribution of URM buildings in LMA according to Parameter P3 indicate that terraced buildings (Class A) account for about 23% of the total number of buildings; isolated buildings (Class B) represent 38%; "Classic" buildings (Class C) account for 19%; and semi-detached or edge buildings account for 17% of the URM buildings in LMA.

Parameter P4 refers to the number of stories (i.e., the height of the building), which affects the seismic vulnerability of the building in terms of drift at roof level and also about the potential pounding effect with adjacent buildings previously also mentioned in the case of parameter P3. Hence, the seismic vulnerability of a building typically increases with the additional number of stories. About 50% of the URM building stock in LMA are low rise (1 floor), with 40% being mid-rise (2–3 floors), and the remaining 10% representing high-rise buildings (above four floors).

Last but not least, parameter P5 evaluates the plan irregularity at the ground floor level when the wall layout changes to create open-plan areas for commercial use. In the case of the URM building typology, a typical example case is the replacement of the front façade masonry walls with full-height glazing windows, which creates a potential eccentricity and torsional effect, increasing the in-plane shear demand on the remaining masonry walls at the ground floor level. Consequently, this type of structural irregularity is assigned as the most vulnerable in class D for the URM building typology. It should be noted that the non-evaluated portion of the building stock has also been ranked conservatively in the most vulnerable class, considering the significant number of buildings included and the uncertainty of what they may represent. On the other hand, Class A represents the least vulnerable cases of regular buildings. The distribution of the URM building typology in LMA for the most and least vulnerable classes previously mentioned shows that class D represents 98% of the URM buildings, of which 16% are irregular and 82% have not been evaluated.

3 Seismic Vulnerability Assessment Results

Based on the methodology described above, once all the seismic vulnerability indices per building have been computed, the results were presented via histogram plots and spatially distributed with the use of the GIS application software (QGIS 3.16.8-Hanover), showing the locations associated with higher seismic vulnerability. Given the large scale of the Census building data, it was necessary to represent the seismic vulnerability results separately for each municipality of the LMA.

3.1 Distribution of the Vulnerability Results for the RC Buildings

The vulnerability index-based method developed in this study was applied to the entire 292,978 RC buildings in the LMA, resulting in a mean seismic vulnerability index of 50.7 and a standard deviation of 15.6. Seismic vulnerability index values range from 11.85 to 93.3. The histogram plotted in Fig. 5 shows that the range of vulnerability 40–50 presents the highest portion of buildings, with 38% of the RC building stock falling within this range. It can also be seen that 22% of the buildings have a vulnerability index within the 50–60 range, about 15% range from 60 to 70, and 11% present a vulnerability index above 70.

3.2 Distribution of the Vulnerability Results for the URM Buildings

The vulnerability index-based method developed in this study was applied to the entire 152,916 URM buildings in the LMA. This resulted in a mean seismic vulnerability index of 37.2 and a standard deviation of 12.6, with vulnerability index values ranging from 14.3 to 91.4. As can be seen in the histogram given in Fig. 6, the range of vulnerability 30–40 presents the highest portion of buildings, 47%. About 19% of

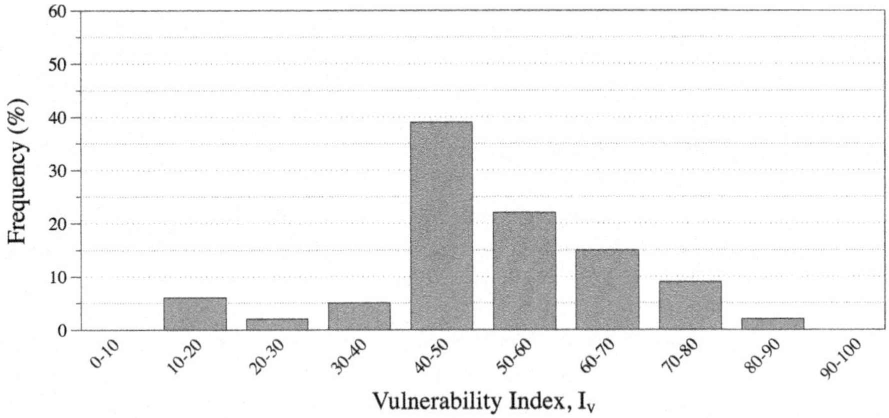

Fig. 5 Vulnerability index distribution of the RC building stock of the LMA

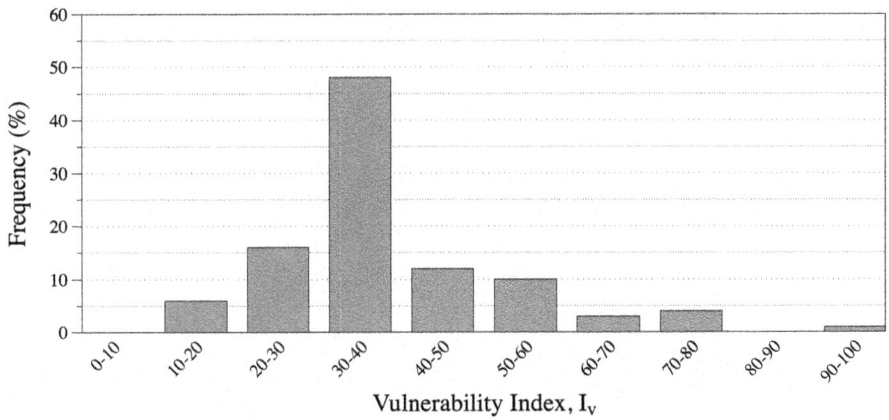

Fig. 6 Vulnerability index distribution of the URM building stock of the LMA

the buildings have a vulnerability index within the 40–50 range, and only 10% are distributed between 50 and 80.

4 Risk Assessment

As mentioned in Sect. 2, this chapter aims to combine seismic vulnerability results obtained using a simplified methodology with seismic hazard maps for the LMA, in order to carry out a seismic risk assessment of the residential building stock of the Lisbon Metropolitan Area.

The seismic vulnerability results and hazard data were manually inputted into a spreadsheet database to create a digital record, which was then implemented and processed using the open-source Geographic Information System software QGIS. Geo-referenced graphical data (i.e., vectorized information and orthophoto maps) and specific information related to the hazard and the characteristics of the buildings were combined within the software to obtain first and second-order outputs.

The integration of the hazard and the vulnerability analysis results in risk outputs that correlate the level of vulnerability with the level of hazard each building is exposed to. The second-order analysis results (i.e., the risk results) of the correlation between vulnerability and hazard levels across the LMA are shown in Fig. 7, illustrating the areas within the LMA most at risk. From this, it can be seen that the higher seismic risk municipalities are mainly Lisboa, Cascais, Amadora, Seixal, and Moita. Something worth noticing from all the above spatial distribution is that areas or municipalities of high to very high hazard and low to moderate vulnerability levels may easily result in a high or very high level of risk, which would not have been expected considering the seismic vulnerability results alone. Typical examples are the Barreiro, Loures, or Setúbal municipalities.

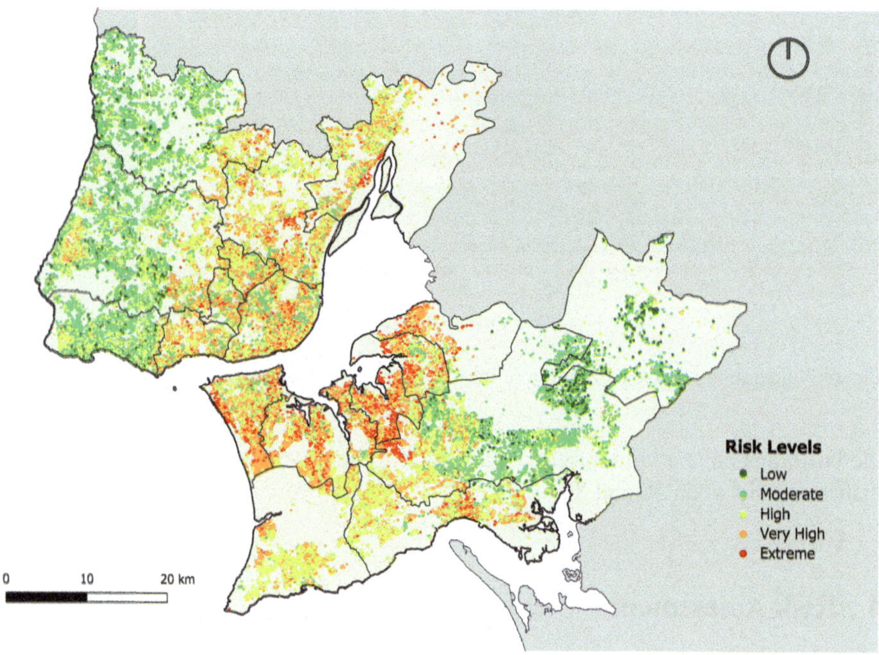

Fig. 7 Spatial distribution of the risk assessment for the residential building stock of Lisbon municipality

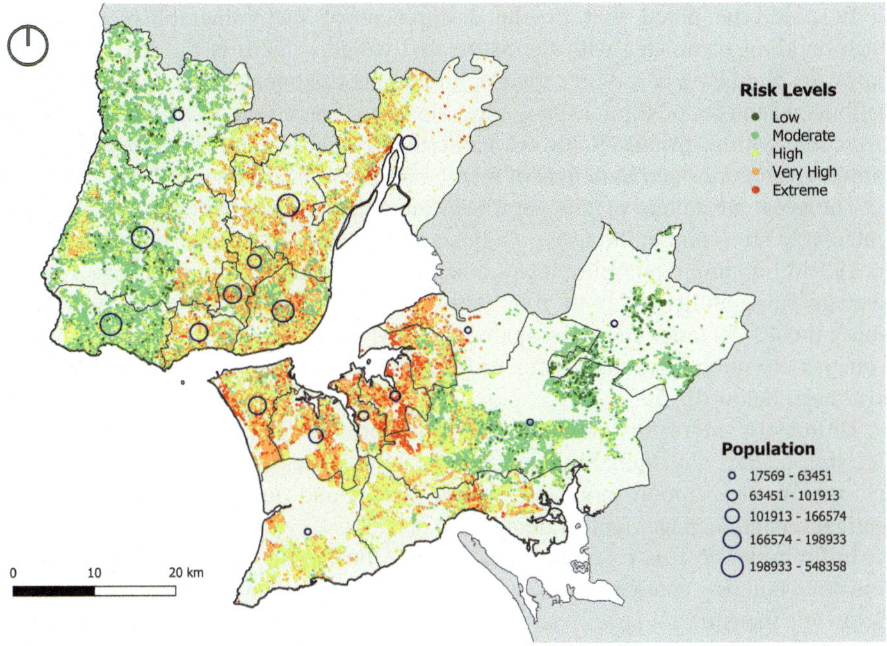

Fig. 8 Spatial distribution of the risk assessment for the residential RC & URM buildings along with the population density over the LMA

Another interesting output of risk analysis is the correlation of risk results with population density as a way to support the development of effective disaster management plans tailored to specific vulnerable buildings and communities—illustrated in Fig. 8.

As can be seen in Fig. 8, the level of risk in Lisbon, which is the most densely populated municipality of the metropolitan region of Lisbon, ranges from moderate to very high or extreme. This should be taken into account when devising mitigation plans for the resilience and preparedness of local communities.

5 Final Remarks

The present work focuses on the seismic vulnerability and risk assessment of the building stock of the Lisbon Metropolitan Area, which encompasses 292,978 reinforced concrete (RC) and 152,916 unreinforced masonry (URM) buildings.

The methodology adopted for the seismic vulnerability assessment suitable for the urban scale is the first level empirical (or statistical) approach using the "Vulnerability Index-based", which uses a considerable amount of qualitative information suitable for large-scale assessment.

It should be noted that on the assignment of the vulnerability classes to each building parameter with the associated weights, there is a degree of uncertainty despite being based on expert opinion, post-earthquake observations, or pre-calibrated samples of similar typologies. However, on certain occasions, this can be overcome with additional fieldwork gathering detailed site information on specific building categories and locations of medium to high-hazard levels.

The great advantage of this approach, though, is the identification of the most vulnerable areas and building typologies within the LMA. Hence, between the RC and URM building typologies, it was concluded that the RC buildings have a higher average vulnerability index of 50.7, with the majority of RC buildings distributed above the 40–50 index range compared to URM buildings, which appear to be less vulnerable with vulnerability index of 37 with the majority being distributed at the low range below 40–50.

Ultimately, the results obtained with this approach can provide technicians and decision-makers with valuable information about the areas at higher risk, which they can then use to support more informed decisions and outline more effective risk mitigation and management strategies.

In future work, it is recommended to refine further any uncertainties regarding unknown building characteristics from the 2011 Census data with carefully planned fieldwork focusing on specifically highly vulnerable and risk areas with the most vulnerable building typologies. For instance, any additional site information for the RC building typologies will refine the vulnerability levels obtained and clarify the classification of the "Classic" and "Non-evaluated" buildings, which represent a high portion regarding the structural irregularity of these buildings. This would enable us to create a more detailed and robust database of the building vulnerabilities and possibly even to re-evaluate the parameters' weights for the vulnerability assessment methodology given the large scale of data provided.

Acknowledgements The project "MIT-RSC—Multi-risk Interactions Towards Resilient and Sustainable Cities" (MIT-EXPL/CS/0018/2019) leading to this work is co-financed by the ERDF—European Regional Development Fund through the Operational Program for Competitiveness and Internationalization—COMPETE 2020, the North Portugal Regional Operational Program—NORTE 2020 and by the Portuguese Foundation for Science and Technology—GCT under the MIT Portugal Program at the 2019 PT call for Exploratory Proposals in "Sustainable Cities".

References

1. Blyth A, Di Napoli A, Parisse F, Namourah Z, Anglade E, Giatreli A-M, Rodrigues H, Ferreira TM (2020) Assessment and mitigation of seismic risk at the urban scale: an application to the historic city center of Leiria, Portugal. Bull Earthq Eng 18:2607–2634. https://doi.org/10.1007/s10518-020-00795-2
2. Custódio S, Dias NA, Carrilho F, Góngora E, Rio I, Marreiros C, Morais I, Alves P, Matias L (2015) Earthquakes in western Iberia: improving the understanding of lithospheric deformation in a slowly deforming region. Geophys J Int 203(1):127–145. https://doi.org/10.1093/gji/ggv285

3. Ferreira TM, Mendes N, Silva R (2019) Multiscale seismic vulnerability assessment and retrofit of existing masonry buildings. Build 9:91. https://doi.org/10.3390/buildings9040091
4. Ferreira TM, Vicente R, Silva JARM, Varum H, Costa A (2013) Seismic vulnerability assessment of historical urban centres: case study of the old city centre in Seixal, Portugal. Bull Earthq Eng 11:1753–1773. https://doi.org/10.1007/s10518-013-9447-2
5. Ferreira TM, Maio R, Vicente R (2017) Seismic vulnerability assessment of the old city centre of Horta, Azores: calibration and application of a seismic vulnerability index method. Bull Earthq Eng 15:2879–2899. https://doi.org/10.1007/s10518-016-0071-9
6. Ferreira TM, Rodrigues H, Vicente R (2020) Seismic vulnerability assessment of existing reinforced concrete buildings in urban centers. Sustainability 12(5):1996. https://doi.org/10.3390/su12051996
7. Ferreira TM, Santos PP (2020) An integrated approach for assessing flood risk in historic city centres. Water 12(6):1648. https://doi.org/10.3390/w12061648
8. Franco G, Shen-Tu B (2009) From 1755 to today—reassessing Lisbon's earthquake risk [Accessed 20 Dec 2021]. https://www.air-worldwide.com/publications/air-currents/from-1755-to-today-reassessing-lisbons-earthquake-risk/
9. GNDT-SSN (1994) Scheda di esposizione e vulnerabilità e di rilevamento danni di primo livello e secondo livello (muratura e cemento armato). Gruppo Nazionale per la Difesa dai Terremoti, Roma, Italy
10. Kassem MM, Mohamed Nazri F, Noroozinejad Farsangi E (2020) The seismic vulnerability assessment methodologies: a state-of-the-art review. Ain Shams Eng J 11:849–864. https://doi.org/10.1016/j.asej.2020.04.001
11. Maio R, Ferreira TM, Vicente R, Estevão J (2016) Seismic vulnerability assessment of historical urban centres: case study of the old city centre of Faro. Portugal. J Risk Res. 19(5):551–580. https://doi.org/10.1080/13669877.2014.988285
12. Maio R, Ferreira TM, Vicente R (2018) A critical discussion on the earthquake risk mitigation of urban cultural heritage assets. Int J Disaster Risk Reduct 27:239–247. https://doi.org/10.1016/j.ijdrr.2017.10.010
13. Montilla JA, Casado CL (2002) Seismic hazard estimate at the Iberian Peninsula. Pure Appl Geophys 159:2699–2713. https://doi.org/10.1007/s00024-002-8754-3
14. Statistics Portugal (2011) Portuguese census survey 2011 [Accessed 30 Oct 2021]. http://censos.ine.pt/
15. Vicente R, Parodi S, Lagomarsino S, Varum H, Silva JARM (2011) Seismic vulnerability and risk assessment: case study of the historic city centre of Coimbra, Portugal. Bull Earthq Eng 9:1067–1096. https://doi.org/10.1007/s10518-010-9233-3
16. Vicente R, D′Ayala D, Ferreira TM, Varum H, Costa A, Silva JARM, Lagomarsino S (2014) Seismic vulnerability and risk assessment of historic masonry buildings. In: Costa A, Guedes J, Varum H (eds) Structural rehabilitation of old buildings. Building pathology and rehabilitation, vol 2. Springer, Berlin, pp 307–348. https://doi.org/10.1007/978-3-642-39686-1_11
17. Vilanova SP, Fonseca JFBD (2004) Seismic hazard impact of the Lower Tagus Valley Fault Zone (SW Iberia). J Seimol. 8(3):331–345. https://doi.org/10.1023/B:JOSE.0000038457.01879.b0
18. Wronna M, Baptista AM, Miranda JM (2021) Reevaluation of the 11 November 1858 earthquake and tsunami in Setúbal: a contribution to the seismic and tsunami hazard assessment in Southwest Iberia. Pure Appl Geophys

Chapter 6
Multi-scale Residential Fire Susceptibility in the Lisbon Metropolitan Area

Carolina Pais, Susana Pereira, and Sérgio Cruz Oliveira

Abstract Urban fires are one the major threats to the security of the urban population. Given their spatial context, they have a direct impact on the economy, inhabitants, and destruction of property and heritage in the community. In the Lisbon metropolitan area (LMA), almost 75% of the urban fires occurred in residential buildings between 2006 and 2020 and were recorded with 150 deaths and 275 severely injured victims. Despite the damages caused by urban fires in LMA, there is a lack of studies on the predisposing factors of residential fire susceptibility and susceptibility zonation at the parish and building levels. This study constructed a residential fire susceptibility model using the information value method, a dataset of socio-economic and building predisposing factors, and an urban fires database for the period 2006–2020. The results suggest that the nearest suburban municipalities from Lisbon record the highest susceptibility to residential fires. Allied with geographic information techniques, these detailed studies are crucial for suited risk reduction strategies and proper urban planning in metropolitan areas.

Keywords Urban fires · Residential fires · Lisbon Metropolitan Area · Fire susceptibility model · Lisbon Metropolitan Area

C. Pais · S. Pereira · S. C. Oliveira
Centre for Geographical Studies, Institute of Geography and Spatial Planning (IGOT), LA TERRA, University of Lisbon, Lisbon, Portugal
e-mail: carolinapais@campus.ul.pt

S. C. Oliveira
e-mail: cruzdeoliveira@campus.ul.pt

S. Pereira (✉)
Faculty of Arts and Humanities, Geography Department, University of Porto, Porto, Portugal
e-mail: sspereira@letras.up.pt

T. M. Ferreira (ed.), *Multi-risk Interactions Towards Resilient and Sustainable Cities*, Advances in Sustainability Science and Technology,
https://doi.org/10.1007/978-981-99-0745-8_6

1 Introduction

Cities face an increased risk of fire, as a result of the combination of high-density buildings and contiguous buildings that facilitate fire propagation and make difficult their extinction by rescue units [1]. According to FEMA [2], urban fires are fires that involve a structure or property within an urban or developed area and are classified according to the building typology where the fire occurs, for example: residential or industrial fire.

An urban fire can have major impacts at the economic, environmental, and social levels, most importantly, the loss of human lives [3]. Given the temporal unpredictability and potential consequences, urban fires are therefore an important hazard in urban areas.

Although an urban fire is not a random occurrence and can be related to certain socio-economic factors, usually poverty [4], likewise, there are other urban fire predisposing factors, such as the building's conservation status and the construction materials associated with the date of construction. In Portugal, this can be seen in older buildings that tend to have more inflammable materials, faulty electrical systems, and lack of maintenance, similar to vacant units [1, 2, 5].

Social demographic factors are equally important to understand the distribution of urban fires [6]. For instance, in Portugal, China, and the USA residential fires are the most common type of urban fires, being more likely to happen in lower-income and higher population density neighbourhoods due to overcrowded units, poor housing, unsupervised children, and improper use of domestic appliances and dangerous substances [2, 7].

In the study of residential fires, most authors use almost exclusively social-economic factors to characterize the spatial fire patterns of a region, mainly because they are commonly available and up to date in official sources, usually in Census data. Therefore, most of the studies have similar conclusions, regarding the variables with the higher predicting power, concerning the spatial patterns of residential fires in Europe, China, Australia, and the USA. Jennings [5], Lizhong et al. [8], Chhetri et al. [9], and Curado [10] found that variables related to population characteristics (high population density, age groups below 16 years old and older than 65 years old, and the presence of indigenous people), socio-economic status (PIB per capita, household income, unemployment, and level of education), and family structure (single parent families) are determinant in the residential fire susceptibility.

Furthermore, an upcoming number of studies have been published, linking urban and residential fires to infrastructure factors. Cumbane and Zêzere [11] showed that there was a higher incidence of urban fires in buildings with more than two floors and buildings with residential, services, and retail typologies. Curado [10] in a study of urban fires in the municipality of Amadora came to consonant results, where buildings with more than two floors, constructed between 1996 and 2000 with metallic and wood infrastructure, were the most susceptible to residential fires.

Normally, urban fire studies focused on infrastructure variables have smaller study areas due to the lack of detailed data and the further need to collect it to produce meaningful analysis. As they produce detailed results, data that was not formerly collected could be easier to gather. Santos et al. [12] evaluated the vulnerability of buildings, in the historical centre of Seixal, regarding infrastructure and surrounding areas. For this study, a similar approach is considered being the industry-type buildings, buildings with faulty electrical systems, buildings without fire detection equipment, and buildings with inadequate evacuation routes classified as the most susceptible.

Despite the importance of urban fires nowadays, most instruments of spatial planning and policy do not acknowledge them, and a few formal governmental studies regarding this hazardous phenomenon have been done such as Vicente et al. [13], Câmara Municipal de Cascais [14], and Zêzere et al. [15]. Consequently, without the proper identification of critical areas, no mitigation policy could be properly implemented.

This present study takes place in the Lisbon Metropolitan Area (Portugal), with 18 municipalities and 118 parishes. According to the population census [15], this is the second most populated area of the country. From 2006 to 2020, there were recorded almost 40 thousand urban fires in the LMA, however, it still lacks a susceptibility map of residential fires for LMA, only a limited number of studies being developed at the municipality, parish, and urban settlement levels: Rocha [16], Zêzere et al. [15], and Curado [10] in Amadora, Santos et al. [12] in Seixal, Nunes [7] in Lisbon. The available database for LMA is very heterogeneous regarding the accuracy and detail of the location of urban fire occurrences. Therefore, it is crucial to evaluate to which extent this database is suitable to assess resident fire susceptibility. According to this context, the main objectives of this study are:

(1) to characterize the temporal evolution and the spatial distribution of the urban fire cases at the parish and municipality levels in the LMA for the period 2006–2020, using the urban fires database obtained from the National Emergency and Civil Protection Authority (ANEPC);
(2) to identify a set of predisposing factors (building characteristics and socio-economic characteristics of the residents) to residential fires;
(3) to assess residential fire susceptibility at a building scale, with a combination of two residential fire susceptibility models regarding socio-economic factors and building factors;
(4) to rank AML parishes and municipalities according to residential fires susceptibility.

2 Study Area

Lisbon metropolitan area (Fig. 1) is a Portuguese region with a population of 2.8 million, living in 1.5 million residential units, and in half a million residential buildings. Almost half of these residential buildings had been built before 1980. Additionally, there are more residential buildings built before 1919 (4%) than residential buildings built between 2001 and 2021 (3.1%). Lisbon municipality (with nearly half a million residential buildings) contributes the most to this statistic. In the peripheral areas, 30% of Barreiro's and Montijo's residential buildings are at least 60 years old, while Alcochete, Mafra, and Sesimbra municipalities have almost 30% of their residential buildings built in the last 20 years [15].

Regarding the population, 32% of them are established in Lisbon (19%) and Sintra (13%) municipalities, although there is higher population density in Amadora (7376.1 inhab/km^2), Lisbon (6389.6 inhab/km^2), and Odivelas (5528.2 inhab/km^2) municipalities. On the opposite, Alcochete, Sesimbra, and Montijo are the least populated municipalities—combining <5% of the LMA population. The age distribution of the population is different depending on the municipality, but it's predominantly mid-age, between 15 and 64 years old. At least 17% of the population of Alcochete, Mafra, Sesimbra, and Sintra are youngsters (<14 years old), and Vila Franca de Xira municipality recorded the lowest percentage of elderlies (more than 65 years old) in the LMA—both 14%. Lisbon has the lowest percentage of youngsters and mid-age percentages—13 and 61%, respectively—but the highest concentration of elderlies (26%).

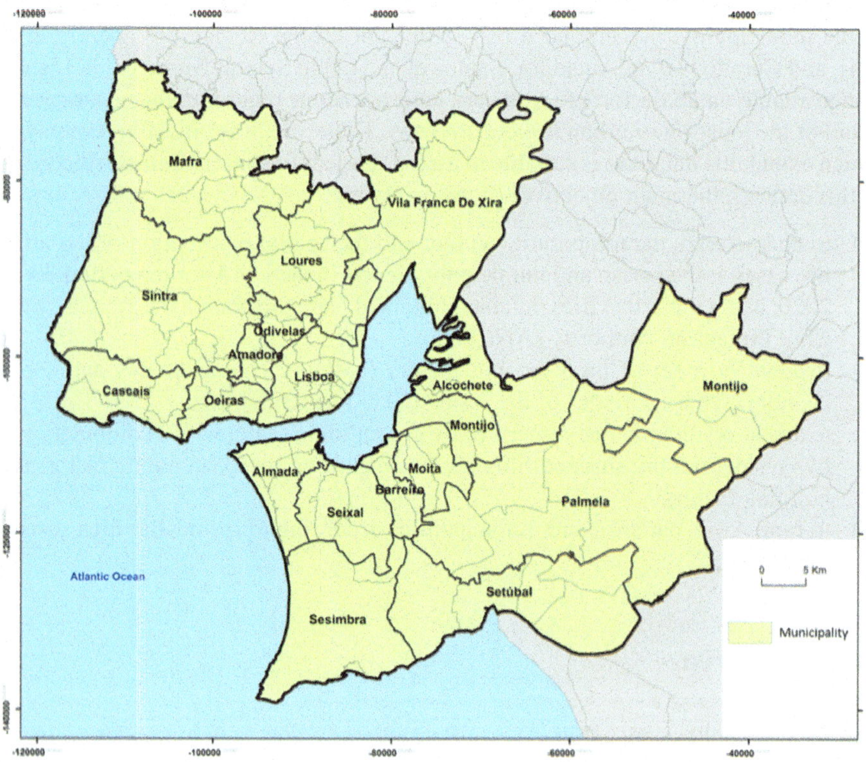

Fig. 1 Municipalities of Lisbon Metropolitan Area

3 Data and Methods

3.1 *Urban Fires Database*

In this study, an urban fires database obtained from the National Emergency and Civil Protection Authority (ANEPC) was used to characterize the temporal evolution and the spatial distribution of the urban fire cases at the parish and municipality level in the LMA for the period 2006–2020. The ANEPC database combines the records of the incidents in which the firefighters were called. The information available includes the geographic coordinates, time (year, month, day, and hour), classification of the dispatch, duration of the fire (in minutes), municipality, parish, building typology (residential, administrative services; army, security forces and emergency unit; dilapidated and vacant building; hospital, health centre, and nursing home; factory, workshop, and warehouse; mall and transport station; parking lot; school and university; show and museum; sports and leisure), victims (fatalities, seriously injured people, injured people, and other victims), and a short description of each event.

The urban fires database provided by the ANEPC includes 36,986 urban fire records in LMA during 2006–2020. For this analysis were not considered incidents classified as false alarms (26%), firefighter dispatches (1.6%), and missing information (0.3%), meaning only 29,039 records were eligible for a detailed analysis.

The description field is the only optional answer and open answer field, therefore the information is miscellaneous and could include information about the source of ignition, transcripts of the emergency call, and full addresses. However, most of the records are blank.

For the geolocation of an urban fire, it was assumed that the coordinates were not accurate after noticing discrepancy patterns. Using satellite images in a GIS environment, each record was associated with the quality of the georeferencing process: (a) building location, if the address was available on the description field; (b) parish's centroid.

Regarding the date of occurrence, two other fields were created: (a) day of the week; (b) day period ("00–03 h", "03–06 h", "06–9 h", "9–12 h", "12–15 h", "15–18 h", "18–21 h", and "21–00 h").

3.2 *Residential Buildings Database and Population Data*

Another crucial database for this study was the Georeferenced Buildings Database (BGE) for the year 2011, obtained through Statistics Portugal (INE). This database only contains information about residential buildings (represented as a point shapefile) and it was used to connect the residential fires to the respective buildings affected and for further analysis. Beyond having the georeferenced 449,573 residential buildings of the LMA, it also has information about the building (number of units, number of floors), structure of building (typology, distribution regarding the surrounding buildings, age, structural material, need of repairs, and type of roof), and the residents (number of residents, gender, age, and number of families).

The demographic information (e.g., age, unemployment, and families) was obtained from the Population Census of 2011 available at the statistical subsection scale which does not match each individual building.

3.3 *Modelling Strategy*

In this study, the purpose was to access the susceptibility of residential fires in residential buildings as a terrain unit, using social-economic factors and building characteristics, applying a bivariate statistical model that incorporates the occurrences of residential fires as a dependent variable. Two susceptibility models were built individually: model (A) building characteristics model; model (B) social-economic model.

To build model A, only the structure of the building variables were considered: (1) type of building; (2) main function; (3) need of repairs; (4) structure's material; (5) type of roof; (6) age; (7) isolated—if the building is isolated or if it is 5 times higher than the contiguous buildings; (8) higher—if it's two storeys higher than the surrounding buildings; (9) contiguity—if the building is detached, semi-detached, or townhouse; (10) exterior cladding of the building. Each variable ranges from 3 to 9 classes by default. These variables were previously used in a LMA study by Curado [10] that modelled the residential fire susceptibility in Amadora.

Model B concerns exclusively social-economic factors. This model was inspired by Jennings [5] that studied the spatial patterns of residential fires in Memphis. The author found that the variables with the higher correlation were (a) vacant units, (b) single-parent families; (c) population with less than 16 years old and more than 65 years old; (d) household income. It was not possible to replicate this model in the LMA because variable (d) it's not available, although other authors used as another option the unemployment percentage because of the high correlation between them [8, 9]. Variable (b) was also switched to residents with less than 19 years old and more than 65 years old, using the demographic information (Census 2011) of the buildings database which was computed with a dasymetric method.

Each variable was calculated at the parish unit (except variable b), reclassified into 4 or 5 classes, and then merged with the dataset of residential buildings. This process enables linking each parish building to the social-economic variables.

From 2006 to 2020, a total of 21,514 residential fires were recorded in LMA but only 74 had a valid address available (RFB). Using Google Earth satellite images in a GIS environment, each record was associated with the correct building (Table 3). This dataset was used to (a) locate the buildings with residential fire occurrences in the LMA, and the corresponding building characteristics; (b) assure contrast in the model. The location of the other 21,440 residential fires reported in the ANEPC database is reported at the parish level, preventing a detailed analysis of the building's characteristics.

3.4 Methods

The method chosen to access the susceptibility of residential fires was the Information Value (IV) (Eq. 1); this bivariate statistical method [17], which allows studying the statistical relation between the dependent and independents variables, is given by Eq. (1):

$$li = ln\frac{Si/Ni}{\mathrm{S/N}} \quad (1)$$

where *Si* is the number of residential buildings where a fire happened in a certain variable class; *Ni* is the number of residential buildings that exist in a certain variable class; S is the total number of residential buildings where a fire happened in the study area; N is the total number of residential buildings in the study area.

The higher the *li*, the higher the possibility that a residential fire happens within a given building. Because of the choice of this method, all data was converted into a raster data structure, so the residential building centroids correspond now to pixels with 10×10 m. The relevance of any independent variable to discriminate susceptible and non-susceptible areas is as greater as its distance from the 0 value of IV. When the score is negative, it means that the presence of the variable Xi is unfavourable to residential fires. Positive scores mean a positive relationship between the presence of the variable and the residential fire occurrence, as high as the higher the score. Information values equal to zero mean no clear relationship between the variable and the residential fire occurrence.

The final scores of the residential fire susceptibility output were divided into 5 classes of susceptibility: (1) Very Low: <-1; (2) Low: between -1 and -0.5; (3) Moderate: between -0.5 and 0.5; (4) High: between 0.5 and 1; (5) Very High: >1.

Despite using only 74 residential fire occurrences, the final models obtained consonant IV scores and variable classes hierarchy similar to Curado [10] that used the same method, dependent variable and building variables, in the municipality of Amadora.

The computation of the area under the curve (AUC) is a metric used to estimate the susceptibility model quality and success-rate curves [18, 19], which has been applied for both models. The AUC value suggests the overall prediction accuracy of the model—there is a higher prediction when it's closest to 1 (100%) [20]. A model is only acceptable if the AUC is higher than 0.75 (75%) and it is considered great when the value is higher than 0.8 (80%) [21].

The spatial agreement between both residential fire susceptibility models was also established using the Map Comparison Kit Software [17], on a cell-by-cell comparison and kappa statistics. In both models, the residential fire susceptibility results were divided into 5 classes of susceptibility, each one with 20% of the study area, ensuring that equivalent susceptibility classes cover the same fraction of the study area in both maps. These results were compared with the calculation of Kappa, that is, the percentage of agreement between two maps, corrected for the fraction of agreement that can be expected by pure chance [22]. If the same susceptibility class was recorded in both models in the same pixel, it would be attributed to the value 1, if not the value is 0.

Finally, both susceptibility models A and B were integrated—Model C—based on the intersection of the susceptibility class area and residential fires area correctly predicted in a contingency table (Table 1).

For the final susceptibility map, the statistical mode of each municipality and parish pixels was used. To ensure the accuracy of Model C, the three models were computed and validated, class by class through the effective ratio. This method is expressed by the ratio between the percentage of residential fires predicted in the class and the percentage of the class area in the study area. The ratio of effectiveness should be either higher than 3 or lower than 0.2 for a prediction class to be significant.

Table 1 Urban and residential fires frequency and consequences from 2006 to 2020, in LMA municipalities

				Residential fires		
Municipality	Urban fires (Nr.)	False alarms (%)	Residential fires (%)	severely injured (Nr.)	Fatalities (Nr.)	Located at the building (Nr.)—RFB
Alcochete	218	28.0	62.8	1	6	0
Almada	2355	19.0	74.1	14	30	21
Amadora	1816	25.9	78.1	13	19	2
Barreiro	949	28.6	70.0	2	4	10
Cascais	2403	33.4	75.2	6	14	1
Lisboa	3021	14.7	75.7	19	21	2
Loures	2180	25.0	73.3	14	16	0
Mafra	813	27.7	65.4	1	9	0
Moita	722	15.1	72.4	5	3	3
Montijo	638	24.1	63.8	4	4	0
Odivelas	1543	23.1	74.7	9	19	5
Oeiras	2001	41.8	75.3	9	8	1
Palmela	769	24.2	68.1	3	4	3
Seixal	1797	19.4	78.1	5	26	9
Sesimbra	492	26.6	74.2	1	8	3
Setúbal	1696	34.6	69.3	5	16	11
Sintra	4277	32.1	77.4	18	24	3
V. F. de Xira	1347	28.2	71.0	4	7	0

To indicate significantly effective classes, the ratios are preferably larger than 6 or <0.1 [23].

4 Results and Discussions

4.1 Urban Fires Characteristics

Between 2006 and 2020, 29,133 urban fires occurred in the LMA distributed by 11 building types. Nearly 75% of the urban fires took place in residential buildings, the other building types less affected. Fires in factories, workshops, and warehouses represent 7.8%, hotels and restaurants have almost 6%, and dilapidated and vacant buildings about 5%.

Only 0.3% of the urban fires could be associated with the respective building, 37.6% with the parish's centroid, and the remainder 62.1% were impossible to verify their accuracy, therefore they were located at the parish's centroid.

Sintra and Lisboa are the most populated municipalities and consequently the ones with more urban fires, with 4,277 and 3,021 urban fires, respectively (Table 1), opposing Alcochete (218) and Sesimbra (492) with less population and urban fires. The average percentage of false alarms is 26%, however, Oeiras has almost 42%, and Cascais, Setúbal, and Sintra have values above 30%. The fires analysed resulted in 8,000 injured people, of which 275 were seriously injured and with a greater representation in Almada (35) and Sintra (31). Lisboa has a higher concentration of fatalities (24) caused by urban fires, followed by Sintra (22). Moreover, Amadora, and Seixal have the higher percentage of residential fires (78.1%) and Alcochete (62.8%) the least.

4.2 Temporal and Spatial Distribution of Urban Fires

The annual distribution of urban fires reveals a descendent trend with two distinct periods (Fig. 2). Before 2012, the annual average number of urban fires was 2,100 with a maximum of 2,187 in 2006. After 2012, the average number of urban fires decreases by more than 8% (1,784). On the monthly scale, the differences between the number of urban fires range from 7 to 10%. There are more incidents in the coldest months, January (9.7%) and December (10.3%), and less in the milder temperature months, meaning April (7.6%) and October (7.4%). Additionally, an in-depth analysis of the ANEPC database allows observing that more urban fires occurred on weekends (30%) and more frequently between 6 and 9 pm (20.6%), during dinner time.

There is a concentration of urban fires surrounding Lisbon municipality, along the Sintra-Amadora-Odivelas-Loures municipality axis (Fig. 3a), although at the parish

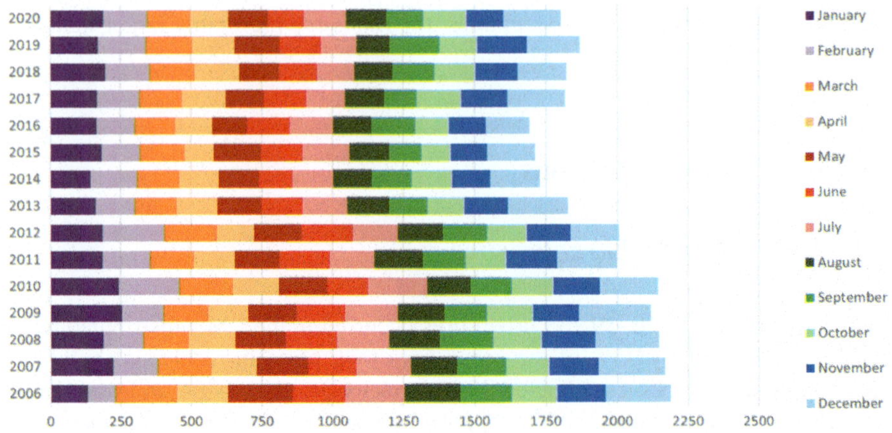

Fig. 2 Annual and monthly distribution of urban fires from 2006 to 2020, in LMA

level the result changes, with a higher concentration in the historic centre of Lisbon, which records more than 80 occurrences per km^2: 58—Santa Maria Maior (142), 19—Campo de Ourique (111), 43—Misericórdia (107), 59—Santo António (105), 12—Arroios (96), and 15—Beato (86). Other parishes with at least 110 occurrences per km^2 are 94—Massamá e Monte Abrãao (146) in Sintra; 45—Odivelas (135) in Odivelas; 96—Moscavide e Portela (135) in Loures; 29—Encosta do Sol (113) in Amadora.

Concerning the residential fires per residential buildings (Fig. 3b), the results show a concentration surrounding Lisbon municipality. The parishes with more than 0.15 residential fires per number of residential buildings are the ones by descending order: 94—Massamá and Monte Abraão (0.24) in Sintra; 109—Santo António dos Cavaleiros and Frielas (0.20) in Loures; 116—Venteira (0.19) in Amadora; 114—Cacém and São Marcos (0.18) in Sintra; 70—Agualva e Mira-Sintra (0.16) in Sintra; and 45—Odivelas (0.15) in Odivelas. This means that in a 14-year period, at least one residential fire happened per seven residential buildings in these parishes. The less hazardous parishes are, for instance: 67—Sesimbra (Castelo) (0.003) in Sesimbra; 79—Azueira and Sobral da Abelheira (0.008) and 28—Encarnação (0.008) in Mafra; 35—Gâmbia-Pontes-Alto da Guerra (0.009) in Setúbal. These parishes are characterized for having a lower population density and consequently a more rural character.

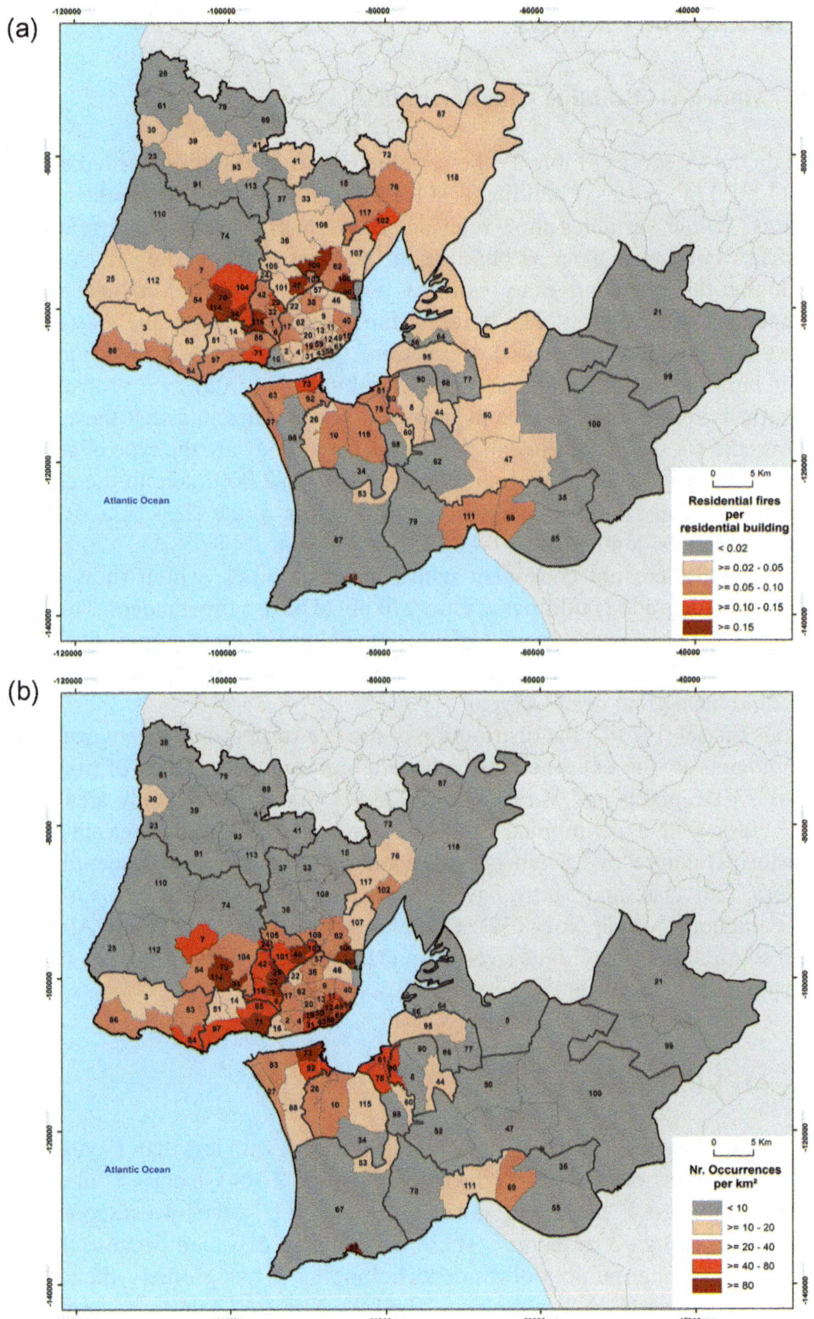

Fig. 3 Spatial distribution of urban fires per parish km^2: **a** and residential fires per residential buildings, **b** from 2006 to 2020, in LMA

4.3 *Suscetibility Models*

4.3.1 Model A—Building Characteristics

Based on the IV scores (Table 2), the variable Higher buildings recorded the highest value (1.801) in the class "building is at least two storeys higher than the contiguous buildings". Other variables show relevant scores that favour the occurrence of residential fire buildings: isolated building in the class "the building is isolated or 5 times higher than the contiguous buildings" (1.327); type of roof in the class "Mixed (Inclined and Terrace)" (1.271); building function in the class "the majority of the building is destined to residential occupation" (1.165); type of building in the class "classic building, built for at least 3 residential units" (1.008).

Contrarily, other classes have negative IV scores, meaning that these are more unfavourable for the occurrence of residential fires. They are the type of structure in the class "brickwork walls without concrete slab" (−2.069); age in the class "built between 1919 and 1945" (−1.588); type of building in the class "classic building, built for 1 or 2 residential units" (−1.425).

Some results are not consonant with previous studies, which show that older buildings and degraded buildings are more likely to have a fire incident. This happens because the variable age does not take into consideration, for example, that a building built between 1919 and 1945 could be renovated and a building between 1961 and 1970 could have great need of repairs [10].

In this model (Fig. 4), the distribution of the fire susceptibility between classes is not homogenous; the extreme classes have a higher concentration of pixels—Very low has 73.1% pixels and Very High has 23.9% pixels of the study area—and the middle classes have a combined value of 3% pixels of the study area. In the LMA, the historical centre of Lisbon is on the intermediate class of susceptibility, and Amadora, Odivelas, and Sintra have several parishes in the highest susceptibility class: Falagueira-Venda Nova (32); Odivelas (45); Massamá e Monte Abraão (94). On the south side of the Tagus river, Almada, Seixal, and Barreiro also show high-dimension building clusters of very high susceptibility.

4.3.2 Model B—Social-Economic

Based on the IV scores (Table 3), the variable residents with less than 19 years old and more than 65 years old show both the highest scores (2.304) in class 5 "more than 15 residents" and class 4 "between 10 and 15 residents" (1.7); the lowest score (−1.012) on class 1 "less than 2 residents". These results were expected because the highest classes have a higher number of residents belonging to risk groups—the elderly and youngsters. The variable "unemployed" had the expected results, meaning that the higher scores belong to classes with the highest percentage of unemployment and vice versa. Controversially, the variables "vacant units" and "single parent families"

Table 2 Building characteristic variables and corresponding information value scores

Code	Variable and classes	S	N	Si	Ni	IV
Type of building						
1	Classic building, built to have 1 or 2 family rooms, Detached	74	449,573	11	175,462	−0.965
2	Classic building, built to have 1 or 2 family rooms, Germinated	74	449,573	6	71,358	−0.672
3	Classic building, built to have 1 or 2 family rooms, in townhouse	74	449,573	3	75,783	−1.425
4	Classic building, built to have 3 or more family dwellings	74	449,573	53	117,413	1.009
5	Classic building of another type	74	449,573	1	8940	−0.386
9	Another type of inhabited building	74	449,573	0	617	−1.426
Main function						
1	Building exclusively for housing	74	449,573	53	407,077	−0.234
2	Building with most of the area allocated to housing	74	449,573	20	37,897	1.165
3	Building with most of the area allocated to purposes other than housing	74	449,573	1	3982	0.4224
9	Not applicable	74	449,573	0	617	−0.235
Need of repairs						
1	Does not require repairs	74	449,573	40	306,737	−0.030
2	Requires minor repairs	74	49,573	20	87,779	0.325
3	Requires medium repairs	74	49,573	4	33,957	−0.335
4	Requires large repairs	74	449,573	1	12,960	−0.758
5	Requires major repairs	74	49,573	0	7523	−0.757
9	Non-classical buildings	74	49,573	0	617	−0.757
Structure's material						
1	Reinforced concrete	74	49,573	61	292,978	0.235
2	Masonry walls with concrete elements	74	49,573	11	97,115	−0.374
3	Masonry walls without concrete elements	74	49,573	1	48,138	−2.067
4	Walls of adobe or loose stone masonry	74	49,573	0	7663	−2.066
5	Other (wood or metallic)	74	49,573	1	3062	0.685
9	Not applicable	74	49,573	0	617	−2.066
Type of roof						
1	Sloped roof—Coated with ceramic or concrete tiles	74	449,573	60	410,689	−0.119
2	Sloped roof—Coated with another material	74	49,573	1	11,592	−0.646
3	On terrace	74	49,573	7	16,453	0.950
4	Mixed (Slope and terrace)	74	49,573	6	10,222	1.271

(continued)

Table 2 (continued)

Code	Variable and classes	S	N	Si	Ni	IV
9	Not applicable	74	49,573	0	617	−0.120
Age						
1	Before 1919	74	449,573	0	22,297	−1.589
2	1919–1945	74	49,573	1	28,954	−1.588
3	1946–1960	74	49,573	5	54,006	−0.602
4	1961–1970	74	449,573	14	59,963	0.392
5	1971–1980	74	49,573	16	83,916	0.181
6	1981–1990	74	49,573	16	71,920	0.275
7	1991–1995	74	449,573	7	32,031	0.257
8	1996–2000	74	49,573	7	35,452	0.155
9	2001–2005	74	49,573	6	35,446	0.001
10	2006–2011	74	49,573	2	24,971	−0.747
99	Non-classical buildings	74	49,573	0	617	−1.589
Isolated						
1	Yes	74	449,573	10	16,113	1.327
2	No	74	49,573	43	101,300	0.947
9	Not applicable	74	49,573	21	332,160	−0.957
Higher buildings						
1	Yes	74	449,573	4	4011	1.801
2	No	74	49,573	39	97,289	0.890
9	Not applicable	74	49,573	31	348,273	−0.615
Contiguity						
1	Yes	74	449,573	22	52,155	0.941
2	No	74	49,573	21	49,145	0.954
9	Not applicable	74	49,573	31	348,273	−0.615
Exterior cladding						
1	Traditional plaster	74	449,573	69	421,662	−0.006
2	Natural stone	74	49,573	3	11,515	0.459
3	Tile, ceramic tile, or mosaic	74	49,573	2	12,585	−0.035
4	Others (wood, glass, etc.)	74	49,573	0	3194	−0.005
9	Plaster	74	49,573	0	617	−0.005

scored the lowest in the classes with higher percentages, revealing opposite results than studies by Jennings [5], Chhetri [9], Chandler [24].

In this model (Table 4), 65% of the pixels were classified as very low or low susceptibility, and nearly 19% of the pixels were classified as being high or very high susceptible. The distribution of pixels is more homogenous than in model A and the mid-class has 17% of the pixels. Visually, this model (Fig. 5) reveals similar

Table 3 Social-economic variables and corresponding information value scores

Code	Variable and classes	S	N	Si	Ni	IV
Unemployment (%)						
1	<6	74	449,573	0	13,052	−0.685
2	>= 6–8	74	449,573	9	76,295	−0.336
3	>= 8–10	74	449,573	17	204,278	−0.684
4	>= 10–12	74	449,573	39	132,624	0.578
5	>= 12	74	449,573	9	23,324	0.850
Vacant units (%)						
1	<8	74	449,573	11	40,189	0.506
2	>= 8–12	74	449,573	35	178,955	0.170
3	>= 12–16	74	449,573	21	152,892	−0.183
4	>= 16	74	449,573	7	77,537	−0.603
Single-parent families (%)						
1	<10	74	449,573	4	65,016	−0.987
2	>= 10–12	74	449,573	8	115,296	−0.866
3	>= 12–14	74	449,573	36	128,768	0.527
4	>= 14–16	74	449,573	20	111,950	0.079
5	>= 16	74	449,573	6	28,543	0.241
Residents with <19 years old and >65 years old (Nr.)						
1	<2	74	449,573	20	333,577	−1.012
2	>= 2–5	74	449,573	10	48,428	0.224
3	>= 5–10	74	449,573	20	47,534	0.936
4	>= 10–15	74	449,573	11	12,172	1.701
5	>= 15	74	449,573	13	7862	2.305

spatial patterns to model A, but the Lisbon historical area is classified as being less susceptible.

4.4 *Validation and Integration of the Models*

Both models were validated, and the AUC recorded values of 0.78 in Model A and 0.81 in Model B (Fig. 6), characterized as a satisfactory and very satisfactory model, respectively, according to Guzzetti et al. (2006). In both models, with just 40% of the study area, it was possible to predict almost 80% of the residential fire occurrences.

Although the value of Kappa represented <20% of concordance, between both models, there is only a moderate concordance in the extreme classes—very low and very high—and the intermediate classes have values close to 0. Spatially, the models are concordant in 50% of LMA's parishes, and in Alcochete, Palmela, Sesimbra, and

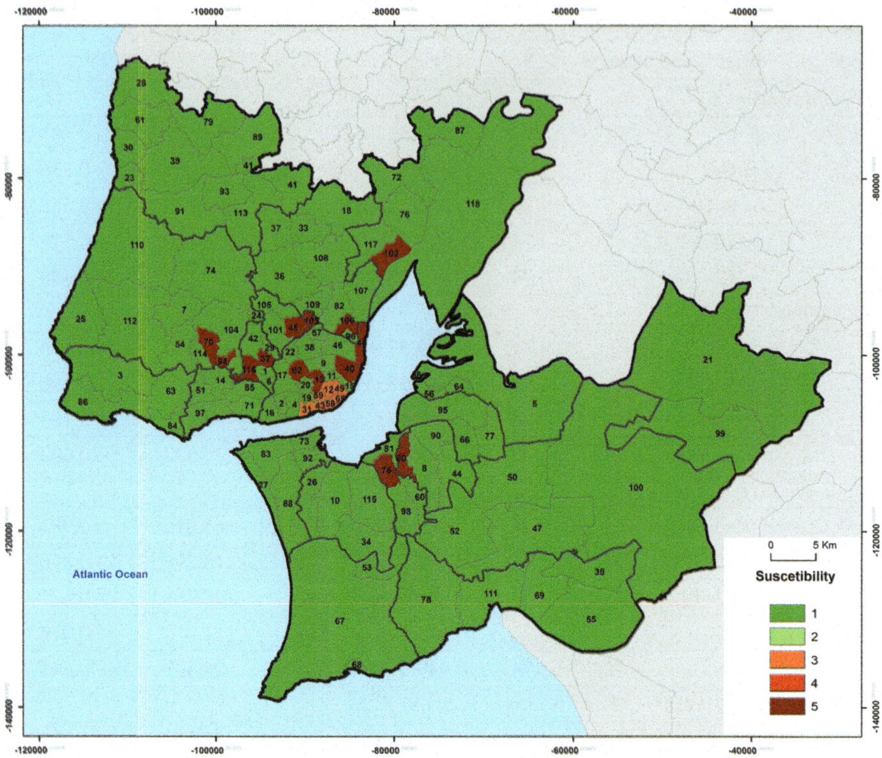

Fig. 4 Residential fire suscetibility results obtained in Model A—building characteristics

Table 4 Distribution of the susceptibility classes between models A and B

Class	Susceptibility	Study area (%)	Study area (%)
		Model A	Model B
<1	Very high	23.9	13.1
1–0.5	High	0.8	5.7
0.5 to −0.5	Moderate	1.6	16.7
−0.5 to −1	Low	0.6	11.6
> −1	Very low	73.1	52.9

Setúbal there is <50% of concordance. The parishes in Lisbon historic centre—São Vicente (65), Santa Maria Maior (58), Misericórdia (43) e, and Santo António (59)—show a concordance below 50%. In opposition, Venteira (116) in Amadora recorded 92% of concordance (Fig. 7). The discordance between the two models justifies their integration. Model C guarantees the complexity of the final susceptibility model to have both social-economic variables and building characteristics. The value "uncertain" (Table 5) was applied when one of the models attributes a higher susceptibility

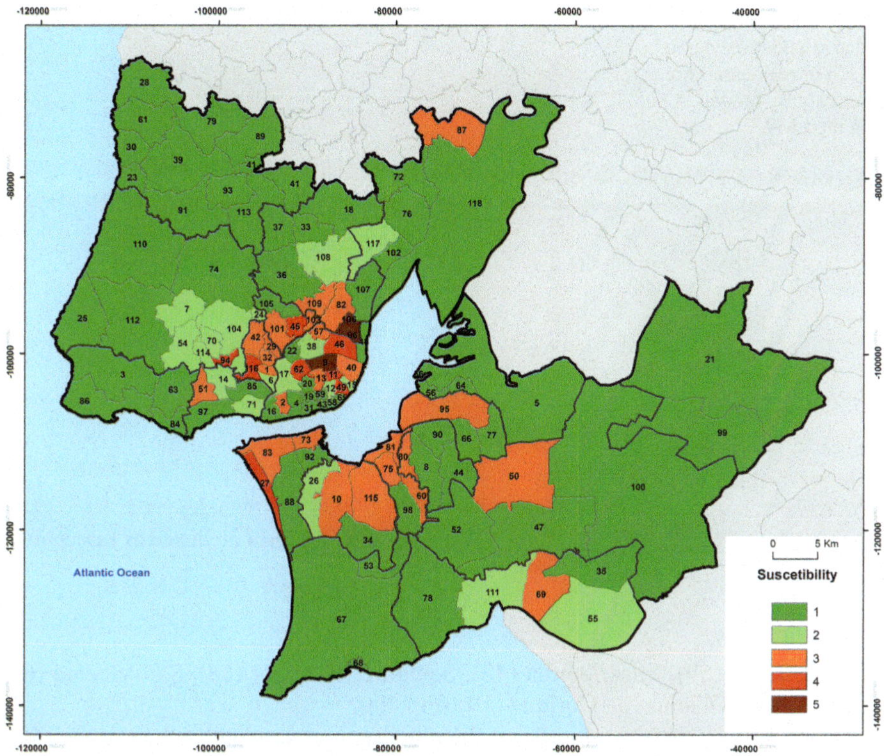

Fig. 5 Residential fire suscetibility results obtained in model B—social-economic

class to a certain pixel and the other model attributes a lower susceptibility class to the same pixel.

There is a higher susceptibility surrounding Lisbon—north and south sides—and in Lisbon itself—Oeiras, Amadora, Odivelas, Lisbon, Barreiro, and Moita are classified as the most susceptible municipalities. Montijo and Setúbal are classified as "uncertain but presumably with high or very high susceptibility", while the other 10 municipalities are classified as having low susceptibility. On the AML north side, the highest concentrations of very high susceptibility parishes are all Amadora's parishes; Oeiras e São Julião da Barra, Paço de Arcos e Caxias (97), Algés, Linda-A-Velha e Cruz Quebrada-Dafundo (71) and Carnaxide e Queijas (85) in Oeiras; Queluz e Belas (104), Agualva e Mira-Sintra (70), Massamá e Monte Abraão (94) and Cacém e São Marcos (114) in Sintra; Odivelas (45) and Póvoa de Santo Adrião e Olival Basto (103) in Odivelas; Santo António dos Cavaleiros e Frielas (109), Sacavém e Prior Velho (107) and Moscavide e Portela (96) in Loures; Povóa de Santa Iria e Forte da Casa (102) in Vila Franca de Xira; in Lisbon, the majority of the parishes are classified as having the higher susceptibility, except near the coastline and/or in the historical centre there are a few uncertain classifications—Ajuda (2),

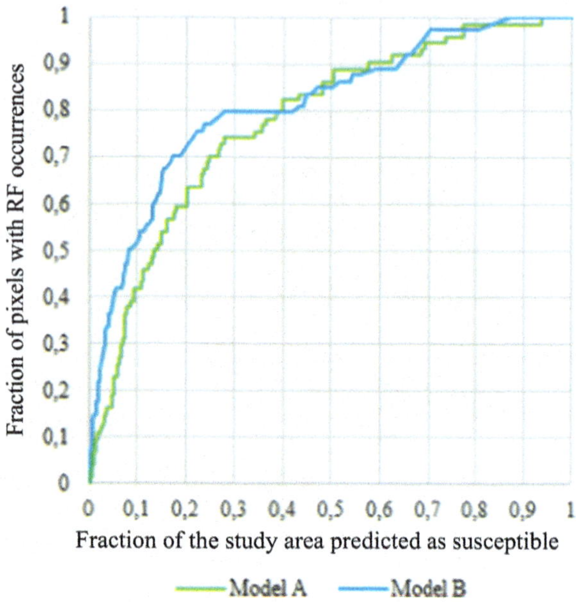

Fig. 6 Sucess-rate curves and area under the curve (AUC) of residential fire susceptibility models A and B in the LMA

Santo António (59), Misericórdia (43), Santa Maria Maior (58), São Vicente (65), Beato (15), and Campo de Ourique (19) (Fig. 8a).

On the south side, the majority of the parishes are classified as low or very low, although near the coastline the scenario changes. The parishes with the higher classifications are Almada, Cova da Piedade, Pragal e Cacilhas (73), Laranjeiro e Feijó (92), and Costa da Caparica (27) in Almada; Alto do Seixalinho, Santo André e Verdena (75) in Barreiro; Baixa da Banheira and Vale da Amoreira (80) in Moita. Other parishes that have high or very high susceptibility classifications but are classified as uncertain are Caparica and Trafaria (83) in Almada; Amora (10) in Seixal; Barreiro and Lavradio (81) in Barreiro; Montijo and Afonseiro (95) in Montijo; Pinhal Novo (50) in Palmela; Setúbal (69) in Setúbal. Mafra, Sesimbra, and Alcochete are the only municipalities where all of the parishes are classified as low or very low susceptible.

Campo de Ourique (Fig. 8b) is one of the parishes classified as highly susceptible to residential fires and also one of the few parishes that has a representation of all of the susceptibility classes. Characterized by having the majority of buildings built within the same period, with similar infrastructure characteristics, and built to have 3 or more family dwellings, these types of buildings are classified with the maximum susceptibility, whereas buildings built to have 1 or 2 family dwellings and buildings built with exclusively housing purposes are classified as being the less susceptible.

The comparison of the effective ratio values, as a means to identifying the value of this integration, revealed a better predictive power and enabled the identification of pixels with different classifications in models A and B. Although the results revealed that the intermediate susceptible classes in models A–C and the highest susceptibility

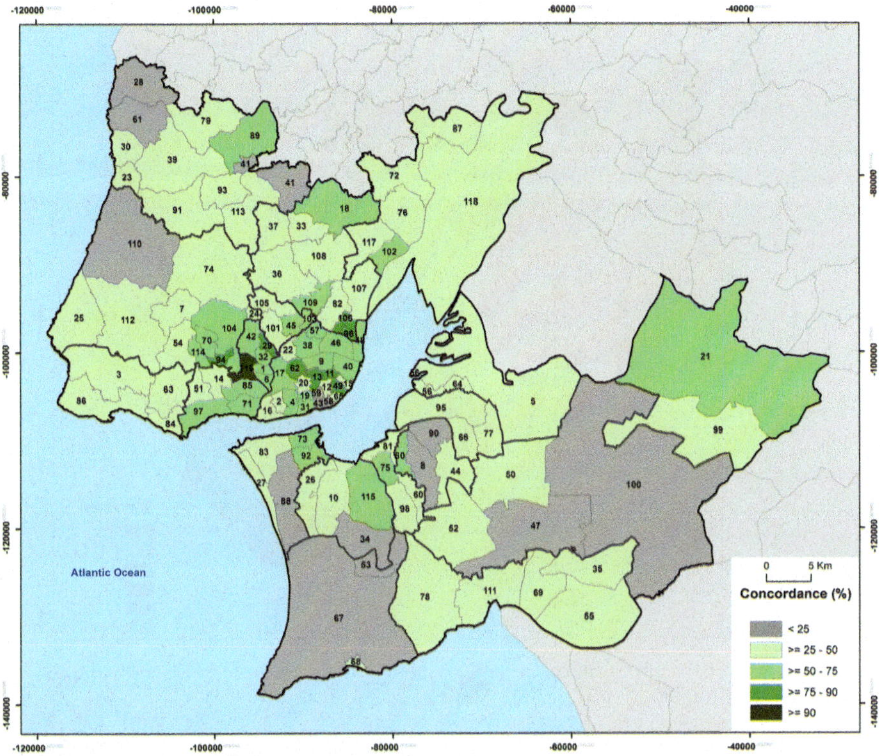

Fig. 7 Concordance (%) between Models A and B

Table 5 Contingency table [25]

Susceptibility	1	2	3	4	5
1	1	2	2	Uncertain	Uncertain
2	2	2	3	Uncertain	Uncertain
3	2	3	3	4	4
4	Uncertain	Uncertain	4	4	5
5	Uncertain	Uncertain	4	5	5

class in model A do not provide enough information to successfully validate the efficiency of the model, the highest and lowest susceptibility class in Models B and C are the only ones that provided accurate validation, with values higher than 3 and lower than 0.2. Nevertheless, the higher susceptibility classes recorded the highest values and were successfully validated—except the "uncertain" class.

The results of Model C admit that 87% of the 74 residential fires are classified in pixels with high or very high susceptibility, and 51% of the pixels are also classified in these susceptibility classes. Comparatively to Model C, Model B gathered 80% of the

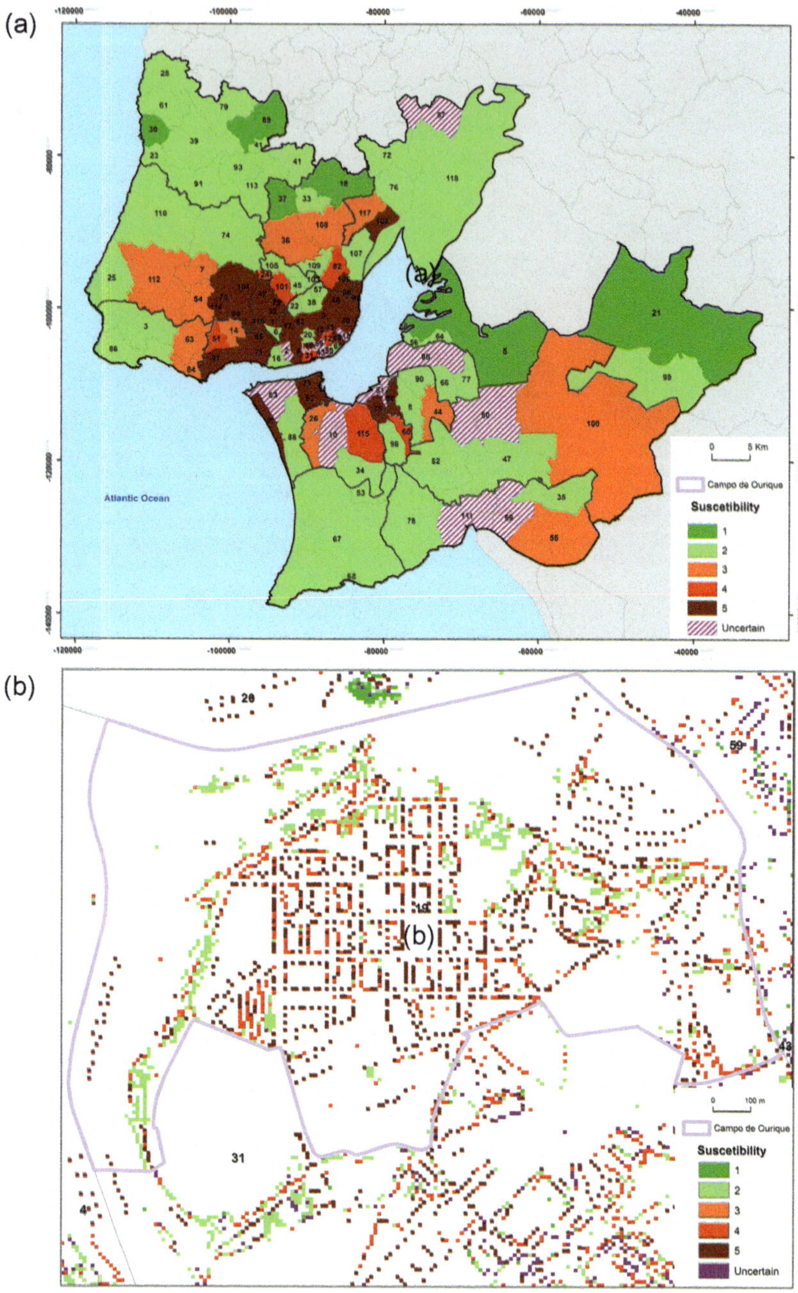

Fig. 8 Residential fire suscetibiliy per parish, in LMA (model C) (**a**) and residential fire sucetibility in Campo de Ourique (model C) (**b**)

Table 6 Effective value comparison between models A–C

Class	Pixel per class (Nr.)	Pixels with residential fires (Nr.)	Study area (%)	Pixels with residential fires (%)	Effective ratio
Model A					
1	84,492	1	20	1.3	0.065
2	84,384	6	20	8.1	0.405
3	84,299	6	20	8.1	0.405
4	84,314	17	20	23	1.15
5	84,179	44	20	59.5	2.975
Model B					
1	87,984	2	20.9	2.7	0.129
2	83,900	6	19.9	8.1	0.407
3	92,881	7	22	9.45	0.430
4	80,828	7	19.2	9.45	0.492
5	76,075	52	18	70.3	3.906
Model C					
1	45,276	1	10.7	1.4	0.131
2	113,302	6	26.9	8.1	0.301
3	42,340	2	10	2.7	0.27
4	60,599	8	14.4	10.8	0.75
5	87,034	52	20.7	70.2	3.391
Uncertain	731,170	5	17.3	6.8	0.3930

residential fires in 37% of the pixels, while Model A gathered 83% of the residential fires in 40% of the pixels of the study area. In this model, 37% of the pixels and 9% of the residential fires are classified as less susceptible pixels (Table 6).

5 Final Remarks

While the body of knowledge has been increasing in the last years, the quality and the availability of data are still the biggest obstacles to assessing residential fire susceptibility at the regional and building scales. Besides the socio-economic information available in official statistics, other data about the environmental and buildings is limited and not frequently updated. The urban fires database also despises any detail, lacking the accuracy of the location in more than 99% of the occurrences and not even providing the addresses. Not only further analysis becomes laborious, but the results could also be too broad and additionally meaningless to access hazard in smaller communities, such as neighbourhoods. Currently, there is no available cartography or studies accessing the susceptibility of urban fires in LMA, and only 1 out of 18 municipalities was formally studied.

This study explored an urban fires database in LMA, Portugal, for a 15-year period. It was possible to demonstrate the spatial distribution and temporal evolution of urban fires, at the municipality and parish scales. The results suggest that there are more urban fires in the north side of the LMA, especially residential fires in the suburban municipalities of Sintra, Amadora, and Odivelas. In addition, the most frequent urban fires occurred in the colder months, ignited by an electrical source, on weekends at dinner time. Although the results are valid, there is a lack of accuracy in the geographic coordinates of the urban fires, which does not allow the identification of hotspots at a neighbourhood level.

The application of the information value was efficient to build the residential fire susceptibility models, given the fact that it measures the relationship between the actual occurrences and characteristics of the population and of the buildings. With this approach, it was possible to determine the most important predisposed characteristics and spatial patterns throughout LMA; and compare each model at the pixel level to reveal the accuracy between both models. Even with only 74 occurrences, the VI scores were consonant and validated with Curado [10] study that studied georeferenced occurrences of residential fires in Amadora, with the same building variables.

Despite the susceptibility models having a very satisfactory predicting power, the majority of the pixels had different susceptibility classifications between both models, resulting in a weak concordance. These models measure different parameters, whereas Model B has reached a higher predicting power because of the implementation of a variable regarding age risk groups. The number of residents and the concentration of younger and older residents are often the best indicators to measure fire susceptibility, given the fact that the misuse of appliances and neglect are the most frequent causes of residential fires [10]. Likewise, these indicators overshadow other indicators, such as concerning building characteristics, and hide other spatial patterns. The integration of both susceptibility models was crucial to diminish this problem and consequently, raise the accuracy of the model. The effective value results are not satisfactory and recorded a majority of meaningless values—even in model C—although the highest classes of susceptibility recorded the highest effective values and the only ones that are valid (except in model A). The results may have been more satisfactory with a higher number of residential fire occurrences, but the pixels classified with the highest and lowest susceptible classes were successfully validated.

Consonant with the spatial pattern analysis of urban fires in LMA, the municipalities and parishes surrounding Lisbon, and Lisbon city, were classified as the most susceptible to residential urban fires. These municipalities have in common the most densely populated in the LMA, such as Lisbon, Sintra, Amadora, and Odivelas.

Regardless of the satisfactory results of this study, there is space for improvement, and future work will be performed to assess all urban fire susceptibilities in the LMA. Given the update of census 2021 information and residential buildings database, it's relevant to update the information to measure potential differences or mitigation policy outcomes. Nevertheless, a need for a more detailed database remains, to improve the accuracy of further studies.

Acknowledgements The project "MIT-RSC—Multi-risk Interactions Towards Resilient and Sustainable Cities" (MIT-EXPL/CS/0018/2019) leading to this work is co-financed by the ERDF—European Regional Development Fund through the Operational Program for Competitiveness and Internationalisation—COMPETE 2020, the North Portugal Regional Operational Program—NORTE 2020, and by the Portuguese Foundation for Science and Technology—GCT under the MIT Portugal Program at the 2019 PT call for Exploratory Proposals in "Sustainable Cities".

References

1. Jennings CR (1999) Socioeconomic characteristics and their relationship to fire incidence: a review of the literature. Fire Technol 35:7–34. https://doi.org/10.1023/A:1015330931387
2. FEMA (1997) Multi Hazard Identification and Assessment, Washington, DC
3. Neto JT, Ferreira T, Remor MB (2020) Análise de Vulnerabilidade ao Incêndio em Núcleos Urbanos Antigos - Os Centros Históricos Brasileiros e a Metodologia ARICA Simplificada. In book: Redução do risco de desastres e a resiliência no meio rural e urbano Publisher: Associação dos Geógrafos Brasileiros, Seção Bauru (AGB/Bauru)
4. Corcoran J, Higgs G, Rohde D, Chhetri P (2011) Investigating the association between weather conditions, calendar events and socio-economic patterns with trends in fire incidence: an Australian case study. J Geogr Syst 13(2):193–226
5. Jennings CR (1996) Urban residential fires: an empirical analysis of building stock and socioeconomic characteristics for Memphis, Tennessee
6. Jennings CR (2013) Social and economic characteristics as determinants of residential fire risk in urban neighborhoods: a review of the literature. Fire Saf J 2013(62):13–19
7. Nunes PMP (2015) Incêndios em edifícios na cidade de Lisboa: Análise e caracterização das ocorrências no período de 2010 a 2014 (Doctoral dissertation)
8. Lizhong Y, Heng C, Yong Y, Tingyong F (2005) The effect of socioeconomic factors on fire in China. J Fire Sci 23(6):451–467
9. Chhetri P, Corcoran J, Stimson RJ, Inbakaran R (2010) Modelling potential Socio-economic determinants of building fires in south east Queensland. Geogr Res 48(1):75–85
10. Curado M (2021) Suscetibilidade a incêndios habitacionais no Município da Amadora. Universidade de Lisboa (Portugal), Tese de Mestrado
11. Cumbane RN, Zêzere J (2018) Susceptibilidade de incêndios urbanos nos bairros do município de Maputo. Revista Científica Academia Ciências Policiais, 19–33
12. Santos C, Ferreira TM, Vicente R, da Silva JRM (2013) Building typologies identification to support risk mitigation at the urban scale—case study of the old city centre of Seixal Portugal. J Cult Herit 14(6):449–463
13. Vicente R, Mendes DS, Varum H, Guimarães DC (2010) Avaliação do Risco de Incêndio Urbano. Cadernos Técnicos de Apoio à Avaliação do Risco Sísmico e de Incêndio nos Núcleos Urbanos Antigos do Seixal. Instituto Pedro Nunes, Coimbra
14. Camara Municipal de Cascais (2015) Dossiê de Suscetibilidades. Disponível em: https://www.cascais.pt/anexo/anexo-21-dossie-de-suscetibilidade
15. Zêzere JL, Reis E, Fonseca N, Garcia R, Oliveira S, Pereira S (2019) Delimitação das áreas de suscetibilidade e de risco no concelho da Amadora. Camara Municipal da Amadora
16. Rocha M (2012) Incêndios Urbanos no Concelho da Amadora, O Risco de Incêndio nas Freguesias da Mina e Venteira. Dissertação de Mestrado em Gestão do Território. Faculdade de Ciências Socias e Humanas. Universidade Nova de Lisboa. Março de 2012

17. Yin KL, Yan TZ (1988) Statistical prediction models for slope instability of metamorphosed rocks. In: Bonnard C (ed) Landslides, Proceedings of the fifth international symposium on landslides, vol 2. Balkema, Rotterdam, pp 1269–1272
18. Guillard C, Zêzere J (2012) Landslide susceptibility assessment and validation in the framework of municipal planning in Portugal: the case of Loures Municipality. Environ manage 50:721–735
19. Zêzere J, Garcia, RAC, Oliveira SC, Reis E (2008) Probabilistic landslide risk analysis considering direct costs in the area north of Lisbon (Portugal). Geomorphology 94(3–4):467–495
20. Wang Q, Guo Y, Li W, He J, Wu Z (2019) Predictive modeling of landslide hazards in Wen County, northwestern China based on information value, weights-of-evidence, and certainty factor. Geomat Nat Haz Risk 10(1):820–835
21. Guzzetti F (2005) Landslide hazard and risk assessment—concepts, methods and tools for the detection and mapping of landslides, for landslides susceptibility zonation and hazard assessment, and for landslide risk evaluation. Universidade de Bona, Dissertação de doutoramento apresentada à Mathematchinaturwissenschaftlichen Fakultät da Rheinischen Friedrich-Wilhelms
22. Visser H, De Nijs T (2006) The map comparison kit. Environ Model Softw 21(3):346–358
23. Chung F, Fabbri AG (2003) Validation of spatial prediction models for landslide hazard mapping. Natural Hazards. 30:451–472
24. Chandler SE, Chapman A, Hollington SJ (1984) Fire incidence, housing and social conditions—the urban situation in Britain. Fire Prevent (London. 1971), (172):15–20
25. Lajas SMA (2016) Integração de métodos estatísticos e determinístico para a avaliação da suscetibilidade a deslizamentos superficiais (Master thesis dissertation)

Chapter 7
On the Physical Vulnerability of Buildings Exposed to Landslide Hazards in the Lisbon Metropolitan Area

Ana Cardoso, Susana Pereira, Tiago Miguel Ferreira, José Luís Zêzere, Raquel Melo, Teresa Vaz, Sérgio Cruz Oliveira, Ricardo A. C. Garcia, Pedro Pinto Santos, and Eusébio Reis

Abstract This study assesses the physical vulnerability of buildings (PVB) in the Lisbon Metropolitan Area (LMA) exposed to landslides triggered by both rainfall and earthquakes. Firstly, a statistical model (Information Value), validated with a ROC curve, was adopted to assess susceptibility to landslides caused by rainfall. Secondly, an Analytic Hierarchy Process was adopted to assess the susceptibility to landslides caused by earthquakes. In this case, the model was validated with an inventory of historical landslides in the LMA. The vulnerability assessment included all

---

A. Cardoso · S. Pereira · J. L. Zêzere · T. Vaz · S. C. Oliveira · R. A. C. Garcia · P. P. Santos · E. Reis
Centre for Geographical Studies, Institute of Geography and Spatial Planning (IGOT), LA TERRA, University of Lisbon, Lisbon, Portugal
e-mail: amcardoso@campus.ul.pt

J. L. Zêzere
e-mail: zezere@igot.ulisboa.pt

T. Vaz
e-mail: tvaz@campus.ul.pt

S. C. Oliveira
e-mail: cruzdeoliveira@campus.ul.pt

R. A. C. Garcia
e-mail: rgarcia@igot.ulisboa.pt

P. P. Santos
e-mail: pmpsantos@campus.ul.pt

E. Reis
e-mail: eusebioreis@campus.ul.pt

S. Pereira (✉)
Faculty of Arts and Humanities, Geography Department, University of Porto, Porto, Portugal
e-mail: sspereira@letras.up.pt

R. Melo
Institute of Earth Sciences, School of Science and Technology, University of Évora, Évora, Portugal
e-mail: raquel.melo@uevora.pt

T. M. Ferreira (ed.), *Multi-risk Interactions Towards Resilient and Sustainable Cities*, Advances in Sustainability Science and Technology,
https://doi.org/10.1007/978-981-99-0745-8_7

residential buildings surveyed in the 2011 Census, considering a set of vulnerability parameters, namely: the presence of reinforced structure, number of floors, conservation status, and need for repairs in the structure and finishes. These parameters, and their respective weights, were based on expert opinion and literature. Through this analysis, it was possible to identify meaningful regional interactions between the earthquake and rainfall-triggered landslides, leading to complex damage scenarios for residential buildings. It was also possible to identify risk hotspots and potential risk adaptation and mitigation measures.

Keywords Landslides · Physical vulnerability · Rainfall-triggered landslides · Earthquake-triggered landslides · Lisbon Metropolitan Area

1 Introduction

Landslides are responsible for major fatalities, generating large human, social and economic losses. Landslides are complex processes with multiple causes, which can occur simultaneously; the final cause is nothing more than a triggering mechanism, which corresponds to the final push of a mass that was already on the verge of rupture. Rainfall and seismic activity are the main triggers of landslides, having already been characterized at the regional scale for some areas in Portugal, namely, the Lisbon region, with studies concerning the susceptibility of rainfall-triggered landslides (RTL) [36, 38]. Works concerning earthquake-triggered landslides (ETL) in Portugal are scarcer since the landslides triggered by this phenomenon are less frequent and older [30].

It is important to consider the context of this work in the current and future framework of the problematic climate situation. Araújo et al. [3], through RCP (Representative Concentration Pathways) emission scenarios: RCP 4.5 and RCP 8.5, which consider the existence of an increase in radiative forcing by 4.5, and 8.5 W/m^2 by the year 2100, respectively, studied and evaluated the impact of extreme precipitation (defined as the amount of precipitation needed to trigger a landslide, rainfall triggering threshold—RTT) on future landslides. This analysis was performed for two locations: Sobral de Monte Agraço (SMA), in the Grande da Pipa river basin, located in the Lisbon district; and Santa Marta de Penaguião (SMP), belonging to the Douro river basin, in the north of the country. For both locations, four thresholds were determined relative to the number of consecutive days with accumulated precipitation: 1D (one day), 10D (10 days), 30D (30 days), and 60D (60 days). The forecast of the RCP4.5 scenario, concerning days with accumulated precipitation mentioned above, shows, in all cases mentioned, an increase in extreme precipitation, between 3 and 10%. In the opposite direction, the forecasts of the amount of future precipitation,

T. M. Ferreira
School of Engineering, College of Arts, Technology and Environment (CATE), University of the West of England (UWE Bristol), Bristol, UK
e-mail: Tiago.Ferreira@uwe.ac.uk

for the RCP8.5 scenario, indicate a decrease in extreme precipitation for 10, 30, and 60 days of accumulated precipitation, for SMA. This suggests that for the RCP4.5 scenario (the most conservative), more extreme precipitation events will occur, with the RTT required for triggering slope movements. For the RCP8.5 scenario, more one-day extreme precipitation events are predicted to occur, isolated episodes with a large concentration of precipitation, with the RTT, required for triggering slope movements, which may lead to an increase in the number of translational surface landslides (associated with intense precipitation events with short duration periods); however, in SMA, the projection of precipitation for 10, 30, and 60 days, has a negative relative difference, which translates to a decrease in the average annual precipitation [19, 24] and a decrease in the occurrence of deeper slope movements (associated with longer precipitation events) [3].

There are several works with records of damage caused by landslides, the most frequent referring to damage caused to population, roads, and buildings, such as the landslides triggered by the seismic event of 1512 (magnitude of 6.3 on the Richter scale) in Lisbon, which destroyed about 200 houses and caused the death of 2,000 people, with a rupture surface between 5 and 7 m [30, 31], and the seismic event of 1597 (magnitude of 5.7 on the Richter scale), which destroyed three streets and 110 houses, with a rupture surface between 15 and 20 m [29, 31].

RTL also causes damage to the population and to different types of structures, buildings, and roads. The DISASTER database identified, in the LMA, for the period from 2006 to 2020, 5 occurrences of landslides movements involving human damage: on October 24, 2006, a rockfall in Almada caused one fatality and, on March 15, 2018, in Colares (Sintra), another rockfall caused one fatality and four injuries. In 2010, there were two mass movements that affected ten buildings, leaving four people homeless [37]. Alves and Oliveira [2] identified 356 landslides that caused damage to roads in the Rio Grande da Pipa basin through the interpretation of aerial photographs and orthophoto maps.

The physical vulnerability of buildings (PVB) refers to the predisposition of the building to suffer damage derived from the occurrence of a potentially destructive natural phenomenon [27] and ranges on a scale from 0 (no damage) to 1 (total damage). According to Oliveira et al. [15], the physical vulnerability of buildings to landslides assessment is based on the following theoretical foundations: (i) the level of damage depends on the typology and magnitude of the landslide; (ii) the level of structural or non-structural damage is largely dependent on the strength of the affected element; (iii) the position of the affected element within the destabilized mass conditions the vulnerability potential; (iv) the level of damage can be qualitatively or quantitatively referenced to determine deformation patterns.

Physical vulnerability assessment methods can be grouped into two types: empirical and analytical methods. Empirical methods consist of the analysis of statistical damage from observed events, or based on expert opinion on physical vulnerability, or the assignment of a score by conducting questionnaires with different parameters to assess the potential damage of different hazardous phenomena [28]. In the case of frequent events, collecting information about the degree of physical damage to buildings or infrastructures is a quick and practical process when using damage probability matrices or vulnerability curves. However, if there is not enough prior information regarding the damage, or if there is no funding for the application of an analytical method, the most viable option is to consult a group of experts specialized in vulnerability or in the expected damage depending on the intensity of the dangerous phenomenon [26, 28].

Analytical methods study the behavior of buildings and structures based on engineering criteria and use physical modeling tests or computer simulation techniques. In these methods, the information about the hazardous phenomenon must be more detailed, which makes these methods more demanding in terms of resources (time and money), but they allow a better understanding of the relationship between the intensity of the hazardous phenomenon and the expected degree of damage to an exposed structure. This factor ends up determining the scale when using the analytical methods since it needs a larger number and greater complexity of data, being used in the assessment of individual structures, in a smaller sample, with a large scale [26, 28].

The building's vulnerability depends on its intrinsic properties (e.g., construction material, conservation state, etc.) and the landslide's dynamic properties (type of movement, speed of movement, volume, etc.). Alves and Oliveira [2] compiled records of the physical vulnerability of different roads (rural road, municipal road; national road; and motorway) in the Grande da Pipa River basin, through a sample of 356 rainfall-triggered landslide movements; and developed an exploratory approach to assess the damage patterns to buildings and roads, resulting from these landslides. Alves [1], assessed the structural vulnerability patterns in roads and buildings generated by the impact of landslides in the same study area, to different internal sectors of the landslide mass: scarp, depletion area, accumulation area, landslide main body, and landslide front.

In this context, this work aims to assess the buildings' physical vulnerability, of the Lisbon metropolitan area (LMA), to landslides triggered by precipitation and seismic activity, individually. For this, the following objectives were defined:

(i) identify the buildings of the LMA that are in very high susceptibility areas to landslides triggered by precipitation and seismic activity;
(ii) create an index of physical vulnerability for the LMA's buildings (BGE), using the variables provided by the 2011 census data and with the support of literature;
(iii) apply the physical vulnerability index to buildings located in the most susceptible areas, either to landslides triggered by precipitation or by seismic activity.

2 Study Area

The LMA contains 18 municipalities (Fig. 1) and 122 parishes, in 3,015.24 km^2, distributed in two main areas, separated by the Tagus estuary: Greater Lisbon and the Setúbal Peninsula. At the national level, the LMA corresponds to the NUT III with the largest number of inhabitants and the second most populous, with about 2,870,770

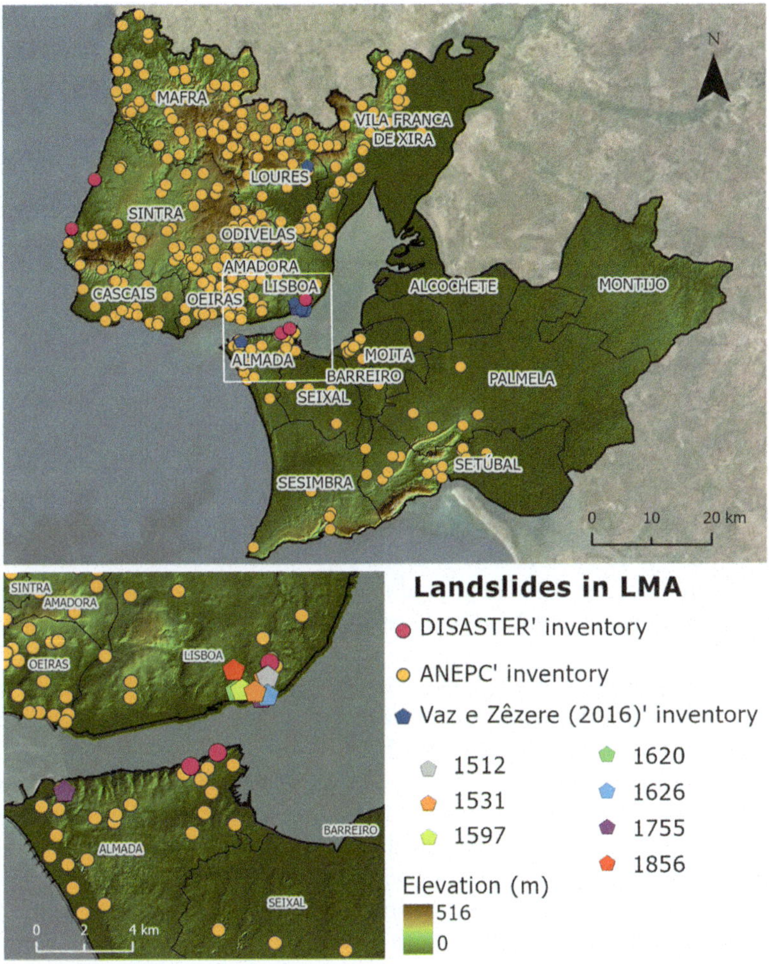

Fig. 1 Landslide inventories triggered by rainfall and landslides in LMA

people registered in the 2021 census, which corresponds to an increase of 1.7% compared to the 2011 census, and represents about 27.8% of the total Portuguese population.

2.1 Geological and Geomorphological Context

The Portuguese territory is located along the Azores-Gibraltar fault, close to the intersection of three tectonic plates, the North American Plate, the Eurasian Plate, and the African Plate [5]. Mainland Portugal is in the southwestern part of the Eurasian

Plate and constitutes an important area of seismic activity, with seismicity resulting either from phenomena derived from movements in faults at the boundaries of the plates (interplates), or from phenomena resulting from faults that are in the interior of the Eurasian plate (intraplates). Earthquakes resulting from interplate movements are typically characterized by high magnitudes and epicentral depths due to the proximity of the continental region to the boundary of the tectonic plates and to the tectonic inversion process, which began in the Quaternary, which translates into important seismic episodes, such as the 1755 earthquake [22]. While earthquakes resulting from intraplate movements have lower and more dispersed seismicity in space, with a maximum magnitude of about 6.5 on the moment magnitude scale [6]. The intersection of the Eurasian Plate with the African Plate, at the southern limit of the Iberian Peninsula, is responsible for the tension that causes the country's high seismic activity, with the converging movement of the two plates in the NNW-SSE direction, translating the subduction of the Atlantic oceanic lithosphere under the Iberian continental lithosphere [25].

Cabral [5] identified different types of faults in Portugal. In the interior and coastal regions, reverse faults predominate, with orientations NE-SW to ENE-WSW and NNE-SSW to N-S, respectively. In the north-east region of mainland Portugal, faults with unplugging movement predominate, with NNE-S orientation, and, on the western Alentejo coast, there are normal faults, namely the Grândola fault, with WNW-ESE orientation, these being the rarest faults type in Portugal. The Lower Tagus Fault is one of the main active tectonic faults in the LMA,characterized by intraplate movement, which movement was responsible for one of the most important earthquakes in the region (January 26, 1531, [31]).

LMA is divided into two morfostructural units: Meso-Cenozoic Basins and the Lower Tagus and Alvalade Cenozoic Basins [20], delimited according to the period of formation and lithological material, which results in a large and complex lithological variety and in the existence of rocks about 260 million years old [21].

On the Meso-Cenozoic Basins, there are two important regional units that are part of the AML territory: the mountains and hills between Montejunto and Lisbon and the Monoclinal Mountains of Boa Viagem and Arrábida [17]. The Mountains and Hills between Montejunto and Lisbon, with an average altitude of 173 m and an average slope of 6.9°, which main relief unit in the LMA is the Serra de Sintra, a dome with 528 m, with an east–west and with an elliptical shape, which results from the ascent of a magmatic massif, and the Lisbon Volcanic Complex, dominated by hills that are what remains of old volcanic cones, deconstructed in the current relief, which have a maximum altitude of 400 m. The Monoclinal Mountains of Boa Viagem and Arrábida present as the main unit of relief in the Arrábida mountain range, with a maximum altitude of 501 m, which stands out in the coastal platform, with a very complex monocline limestone relief, which includes subunits: the Comenda and Vale da Rasca units, the sandstones and transitional clays and the Espichel limestones, sandstones and marls; from east to west, respectively [11].

The Lower Tagus and Alvalade Cenozoic Basins include five regional relief units included in the LMA territory: the High plain of Ribatejo, with average slopes of 5.3°, with a plain situated between 60 and the 100 m, with essentially Miocene sediments, with evidence for the Trancão river basin; the High plain of Alentejo, with an average altitude of 69 m and an average slope of 2°, characterized by the sedimentary filling surfaces; the low Tagus plain, a flat and lower area, with an average altitude of 23 m and an average slope of 1.6°, with alluvium and low sedimentary terraces of the Tagus river; the Peniche-Lisbon Coastal Plain, which corresponds to a narrow coastal sector limited by cliffs and steep slopes, with an average altitude of 70 m and an average slope of 4.6°, and with small or non-existent beaches; and the Setúbal Peninsula, with an average altitude of 43 m and an average slope of 2.3°, located between the Tagus estuary and Serra da Arrábida, whose main relief units are the Belverde platform and the fossil cliff of Costa de Caparica, a steep slope (max. altitude of 100 m) that presents slope instability and soil water erosion.

The characteristics of the varied and complex lithology of the LMA determine the spatial distribution of landslides (Fig. 1) and the way in which the terrain reacts to the passage of seismic waves. Therefore, it is crucial to pay attention to the lithology and the zoning of seismic hazards in urban occupation, in the type of construction, and in the construction of road structures [21].

2.2 *Building Environment in the LMA*

According to 2011 census data, the LMA comprises a total of 448,957 buildings, of which 292,978 (65%) correspond to reinforced concrete buildings, 97,116 (22%) correspond to slab buildings, and 48,138 (11%) correspond to masonry buildings. There was also a small percentage of buildings, around 3%, corresponding to buildings made of adobe or loose stone masonry and other materials (7,663 and 3,062, respectively). The current building includes accommodations that were built before 1919 (22,297). The urban fabric showed an increasing evolution until 1990, and in the 30 years from 1961 to 1990, 215,799 buildings were built, which represents about 48% of current buildings.

Figure 2 illustrates the evolution of the type of building structure used in the construction of the buildings' structure over the years, corroborating the chronological scheme of Bernardo et al. [4], and proving the rise and dominance of reinforced concrete, which started in the mid-1930s. It's also significant the initial decrease and eventual disappearance of materials such as loose stone or adobe masonry walls and the masonry wall without slab from the period 1971–1980, reflecting the transition from traditional building systems and materials to current building technologies.

Moving on to the analysis of the volumetry of the buildings present in the LMA; around 69% of the buildings included in the 2011 census (309,150) correspond to buildings with 1 or 2 floors, where buildings with more than two floors representing around 31% (139,807).

Figure 3 combines the type of structural system of the building with the number of floors. Of the 309,150 buildings with 1 or 2 floors (69% of the total number of buildings), 59.7% correspond to reinforced concrete buildings, 25.2% to slab buildings, 12% to masonry buildings, and only 2.3% to buildings with loose stone

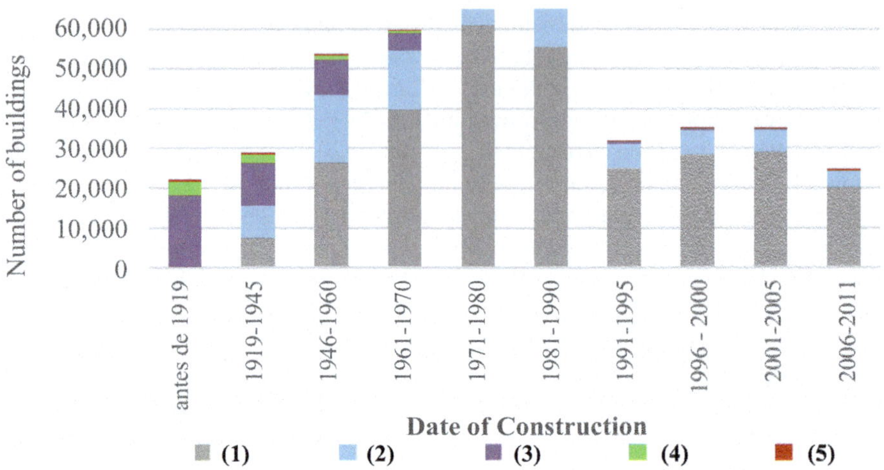

Fig. 2 Type of building structure per date of construction periods in LMA: (1) Reinforced concrete; (2) Masonry wall with concrete elements; (3) Masonry walls without concrete elements; (4) Walls of adobe or loose stone masonry; (5) Others

or adobe masonry walls. The remaining 0.8% corresponds to buildings with 1 or 2 floors made up of other structural systems. Of the 139 807 buildings with two or more floors (31% of all LMA buildings), 77.6% have a reinforced concrete structure, 13.7% are slab buildings, 7.8% are masonry walls and only 0.4% have walls made of loose stone or adobe masonry. As in the case of buildings with 1 or 2 floors, the remaining 0.5% of buildings with two or more floors were built using other structural solutions.

Fig. 3 Reinforced structure and number of floors of buildings in LMA: (1) reinforced concrete; (2) slab buildings; (3) masonry buildings; (4) loose stone or adobe masonry walls; (5) other structural systems

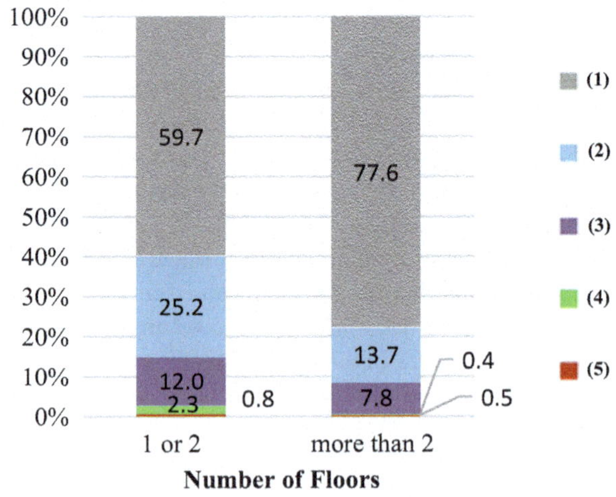

The exterior cladding of buildings, in the LMA, has traditional plaster or marble as its predominant material, with about 93.6% of the total material, dominating during all construction periods. According to INE data from the 2011 census, 91.5% of buildings in the LMA have sloped roofs, covered with ceramic or concrete tiles.

3 Data and Methods

The physical vulnerability of the building located on the slip surface, for landslides triggered by rainfall and seismic activity, is evaluated at the building level, for all LMA buildings, located in very susceptible areas (9th decile) for both triggering mechanisms (Fig. 4).

3.1 Landslide Inventory in the LMA

Landslide inventories were obtained in three different databases: two recorded landslides triggered by rainfall (Disaster database and the National Authority for Emergency and Civil Protection (ANEPC database) and another is a historical inventory of landslides triggered by seismic events. The location of the landslides included in each database can be found in Fig. 1.

DISASTER's national database only includes landslides that caused human damages (e.g., fatalities, injured, evacuated, displaced, and/or missing people) [36] reported in national and regional newspapers. Between 2006 and 2020, there were five damaging landslides in the LMA, where two fatalities, five injured, four displaced people, and ten affected buildings were recorded. The Disaster cases are related to damaging landslides affecting people and/or buildings. An example is the rock fall on cliffs, such as the 2018 occurrence in Colares, which resulted in one death and four injuries [39].

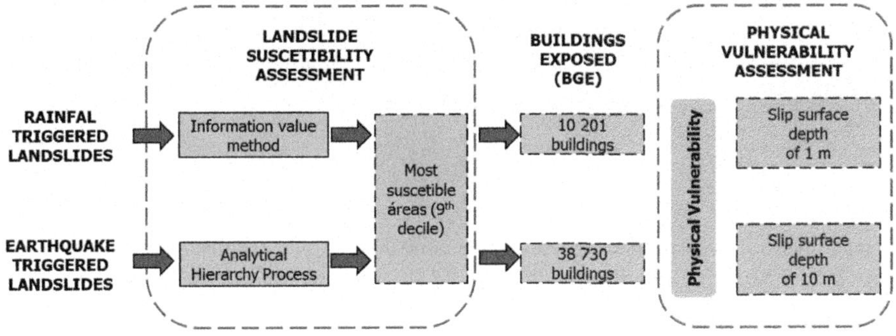

Fig. 4 Methodological framework

The ANEPC database was cleaned in order to remove all records without description and location, reported by civil protection authorities every time there is an emergency call. Seven hundred sixty-six landslides were then identified between 2006 and 2020. The years that recorded the highest number of occurrences were 2010 (154 occurrences), 2009 (91 occurrences), 2006 (77 occurrences), 2014 (76 occurrences), and 2013 (75 occurrences), with Mafra and Loures being the most affected municipalities in these years with 165 and 123 occurrences, respectively. ANEPC database includes more numerous and dispersed landslide cases in the LMA (Fig. 1), often reporting landslides that caused obstructions on public roads and/or damages in buildings.

The inventory of historical landslides triggered by earthquakes identified by Vaz and Zêzere [30] in historical documental sources contains ten occurrences in the LMA between 1512 and 1856. The 1755 earthquake caused the largest number of landslides (3), followed by the 1531 earthquake (2). Santa Maria Maior was the most affected parish, with five landslides triggered by four different earthquakes. The landslides records triggered by earthquakes [30] cannot be dissociated from the date of occurrence, the context of the urban fabric, and the population distribution at the time of the events, since it is in places with a greater population concentration and greater urban density, that these occurrences have more visibility and impact, and therefore more documentary records are available.

3.2 Rainfall-Triggered Landslide Susceptibility

Rainfall-triggered landslide susceptibility was assessed in a previous study [36] using a bivariate statistical model, the Information Value method (IV), using seven predisposing factors with 10 m cell-size (slope angle, slope exposure, slope curvature—transverse profile, geology, land use, topographic position index—TPI and topographic moisture index). Landslides were inventoried for the Lisbon and Tagus Valley region based on the interpretation of aerial photos and orthophoto maps and fieldwork in sample areas. The landslide inventory performed included 4,047 occurrences, most of them corresponding to slides (translational and rotational), which were used as a dependent variable. The information value of each class within each predisposing factor was computed to apply the IV equation, as formulated in Eq. (1) [34, 35]:

$$Ii = \ln\frac{Si/Ni}{S/N} \quad (1)$$

where l*i* is the Information Value of variable *Xi*; *Si* is the number of terrain units with landslides and the presence of variable *Xi*; *Ni* is the number of terrain units with variable X*i*; *S* is the total number of terrain units with landslides; and *N* is the total number of terrain units.

The relevance of any independent variable to discriminate between stable and unstable areas is as greater as its distance from the 0 value of IV. When the score is negative, it means that the presence of the variable X*i* is favorable to slope stability. Positive scores mean a positive relationship between the presence of the variable and the landslide occurrence, as high as the higher the score. Information values equal to zero mean no clear relationship between the variable and the landslide occurrence.

The quality of this rainfall-triggered landslide susceptibility model was assessed through the computation of ROC success rate curves and the corresponding Area Under the Curve (AUC). The rainfall-triggered landslide susceptibility was divided into five classes (very high, high, moderate, low, and very low), defined through the percentage of accumulated landslide area: 50, 70, 90, 95, and 100%, respectively. For delimiting the areas currently subject to the danger of slope instability, high and very high susceptibility classes were selected and aggregated, which together validate 70% of the landslide inventory and, which together, represent the 90th percentile of the susceptible area.

3.3 *Earthquake-Triggered Landslide Susceptibility*

Susceptibility to landslides triggered by earthquakes was assessed with the multicriteria method of the Analytic Hierarchy Process to achieve the relative weights based on Saaty's scale of influence [23]. This technique consists of comparing predisposition factors, ranking one in relation to the other through the attribution of weights (to the variables) and scores (to the classes of variables). This technique is quite subjective as it depends on specialized knowledge of the different predisposing factors to justify the ranking of priority scales [23], in this case, this knowledge is justified with the support of bibliographic research.

There is no universal criterion for choosing predisposing factors, requiring prior technical knowledge of the hazardous phenomenon and the characteristics of the study area, and it is necessary to consider the scale of analysis and data acquisition techniques. In this model, there was used six predisposing factors (slope angle, slope curvature, TPI, geology, PGA, and distance to faults (Fig. 5)) and a historical landslide inventory (10) based on documental sources [30].

Westen et al. [32] present an overview of environmental factors and their relevance to the susceptibility of landslides and hazard assessment, considering the slope as the most important factor for gravitational movements, factors derived from the digital elevation model (DEM) (slope and profile curvature) are of high importance, as are lithology-related factors, structure-related factors, and faults [14, 32]. Slope, lithology, TPI, and curvature are also frequently used factors in susceptibility models performed for study areas located in LMA areas [8, 9].

From the DEM, with a resolution of 10 m, the slope angle, slope curvature, and TPI were generated. The lithology was obtained from the Geological Map of the Lisbon Metropolitan Area, on a 1: 25,000 scale. The five classes of distance to faults were computed with buffers of 1,000 m, 5,000 m, 10,000 m, 20,000 m, and 50,000 m,

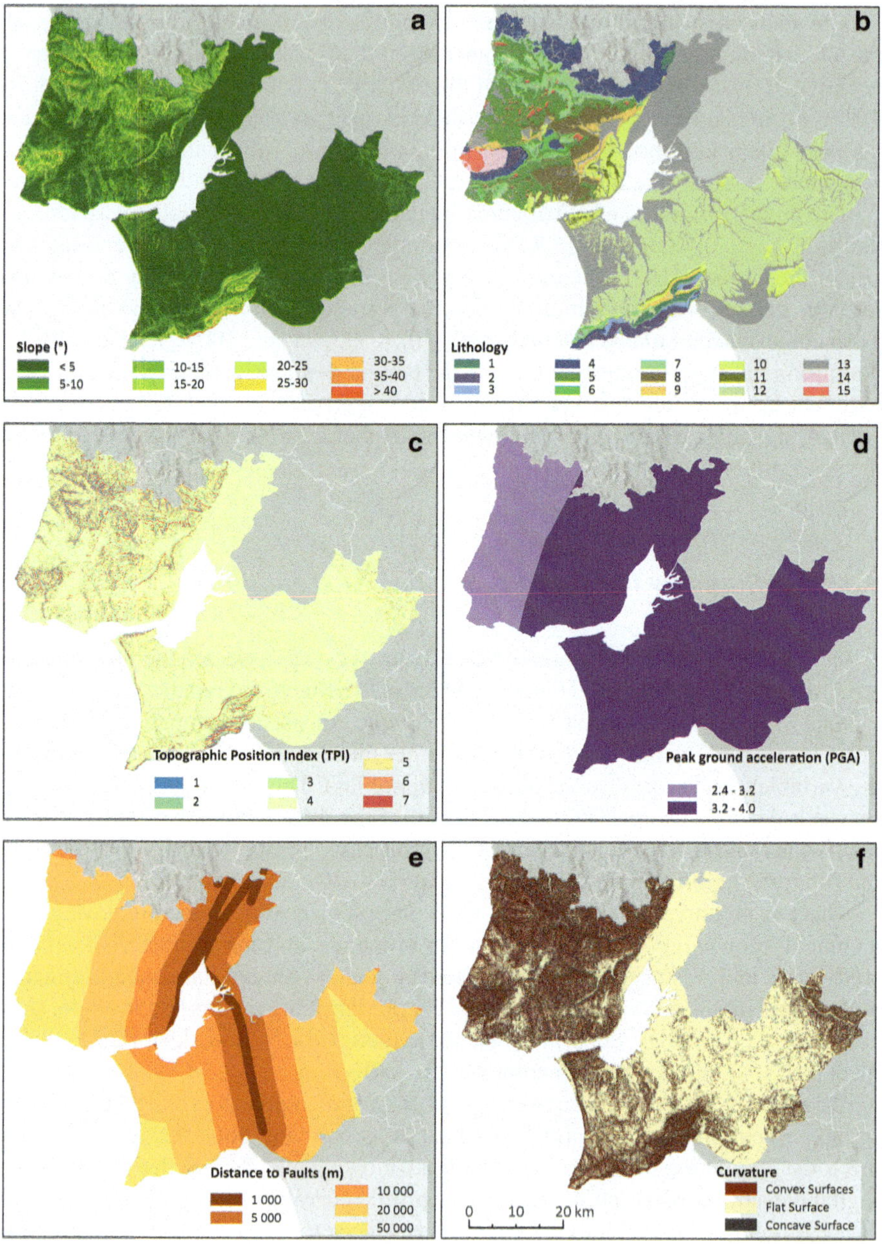

Fig. 5 Predisposing factors and corresponding classes used to assess the earthquake-triggered landslide susceptibility. See legends in Table 2

to faults with tectonic activity provided by the Quaternary Faults Database of Iberia (QAFI). The Peak Ground Acceleration (PGA) was obtained from Peláez Montilla and López Casado [16] for a scenario of peak accelerations with a 10% probability of exceedance in 50 years (return period of 475 years).

The predisposing factors were reclassified, and the respective classes were ranked, by decreasing order of importance, where a weight was assigned to each class and a score to each class of the variable (see Table 2).

The area most susceptible to landslides triggered by earthquakes (9th decile) was selected to assess the PVB for each building exposed to landslides with a slip surface depth of 10 m. A slope movement inventory with ten occurrences [30] was used to assess the overall quality of the susceptibility model of slope movements triggered by seismic activity through the calculation of the success curve and the AUC.

3.4 Physical Vulnerability of Buildings

The physical vulnerability of buildings with a residential function exposed to the most susceptible landslide was assessed for the whole LMA, using the Building Georeferencing Base (BGE) of the Portuguese Institute of Statistics (INE). This database includes 449,473 buildings for the LMA and data about the building characteristics, building type, building structure, main use, number of floors, presence of elevator, accessibility to a wheelchair, number of lodgings, period of construction, the material used in the exterior cladding of the building, type of building cover and the need for repairs (rooftop, structure, walls, and window frames).

In this work, two scenarios were considered: (i) the scenario of a given building is located in a slip surface area of 1 m, in a situation of slope instability triggered by precipitation; and (ii) the scenario of a given building is located in the 10-m slip surface area, in a situation of slope instability triggered by seismic activity. The physical vulnerability index was computed for all buildings located in high and very high susceptibility classes (90th percentile) in both scenarios (10,201 buildings in scenario (i) and 38,730 buildings in scenario (ii)).

The physical vulnerability index includes six parameters: construction material (CM), reinforced structure (RS), construction period (PC), need for repair in the structure (NRS), need for repair in the finishes (NRF), and the number of floors (NF). Each parameter was divided into a set of building classes obtained from BGE (Table 1).

Construction materials (CM) include four main types of buildings from the highest to the lowest resistance: reinforced concrete, brick or stone walls, adobe, and other materials (e.g., wood and metallic). The presence of a reinforced structure (RS) was also considered because, when present, it provides additional resistance to the building. The classes were ranked from the most to the least reinforced structure.

The construction period (CP) in the BGE database includes 11 classes, starting before 1919 until 2011. The construction period indirectly indicates the conservation status and the construction quality of buildings per historical period. This variable was

Table 1 Vulnerability index formulation: parameters, classes and respective weights

Parameters and classes		RTL	ETL
		Slip surface depth 1 m	Slip surface depth 10 m
CM	Reinforced concrete	0.3	0.91
	Brick and stone	0.43	0.99
	Adobe	0.59	1.00
	Other (Wood, metallic, etc.)	0.64	1.00
RS	Reinforced concrete	0.3	0.91
	Masonry walls with concrete elements	0.3	0.91
	Masonry walls. without concrete elements	0.43	0.99
	Walls of adobe or loose stone masonry	0.59	1.00
	Other (Wood, metallic, etc.)	0.64	1.00
CP	>1991	0.3	0.3
	1961 < 1991	0.5	0.5
	1919 < 1961	0.7	0.7
	<1919	0.9	0.9
NRS and NRF	Does not require repairs	0.2	0.2
	Requires minor repairs	0.4	0.4
	Requires medium repairs	0.6	0.6
	Requires large repairs	0.8	0.8
	Requires major repairs	1	1
NF	>2	0.3	0.3
	1 or 2	0.5	0.5

divided into four classes from the lowest to the highest building feature class weight: (i) the period before 1919 to include the most ancient and historical buildings, with adobe/rubble masonry and the unreinforced masonry buildings with timber floor; (ii) the period between 1919 and 1960 to include the two World Wars periods and the shortage of construction materials, such as steel. At this time, there was the emergence of the "Placa" buildings, in Lisbon, constructed with regular geometry; (iii) the period between 1960 and 1990, characterized by the use of better construction techniques, including the use of concrete; and (iv) the period after 1990, when the buildings have a good quality with reinforced construction materials [12, 33].

The need for repair in the structure (NRS) and in the finishes (NRF) includes five classes: "does not require repairs," "requires minor repairs," "requires medium repairs," "requires large repairs," and "requires major repairs."

Table 2 AHP scores for each predisposing factor class to assess earthquake-triggered landslide susceptibility

	ID	Class	Class score	Variable weight
Slope angle	1	<5	0.019	0.460
	2	5–10	0.026	
	3	10–15	0.037	
	4	15–20	0.053	
	5	20–25	0.076	
	6	25–30	0.109	
	7	30–35	0.154	
	8	35–40	0.218	
	9	>40	0.307	
Lithology	1	Sandstones–Cretaceous	0.045	0.256
	2	Sandstones–Jurassic	0.164	
	3	Limestones–Cretaceous	0.264	
	4	Limestones–Jurassic	0.264	
	5	The volcanic complex of Lisbon	0.1	
	6	Volcano-sedimentary complex	0.068	
	7	Quaternary deposits	0.068	
	8	Granite	0.024	
	9	Marls and limestones	0.335	
	10	Marls and limestones–Jurassic	0.335	
	11	Other rocks–Miocene	0.045	
	12	Other acid magmatic rocks	0.045	
	13	Clayey rocks–Miocene	0.068	
	14	Detrital rocks–Paleogene	0.164	
	15	Detrital rocks–Pliocene	0.045	
TPI	1	V-shaped valleys	0.114	0.114
	2	Moderate fit valleys. half slope drainage	0.114	
	3 4	U-shaped valleys Flat areas and slope toe	0.114 0.045	
	4	Flat areas and slope base	0,045	
	5	Middle Slope	0.317	
	6	Upper Slope	0.524	
	7	Top slope	0.045	
PGA	1	3.2–4.0	0.539	0.087
	2	2.4–3.2	0.297	

(continued)

Table 2 (continued)

	ID	Class	Class score	Variable weight
Distance to faults	1	1 000	0.462	0.051
	2	5 000	0.256	
	3	10 000	0.142	
	4	20 000	0.088	
	5	50 000	0.052	
Slope curvature	1	Convex	0.257	0.032
	2	Rectilinear	0.074	
	3	Concave	0.669	

The number of floors (NF) is a proxy variable of the foundations' depth, because the higher the building, the deeper and more resistant the foundations should be Silva and Pereira [26]. In this work, two classes of the number of floors were considered: 1 or 2 floors and more than two floors.

A score was given to each building class and the respective parameter (see Table 1) for both triggering scenarios. Both scores and parameters' weights are based on expert opinion and dedicated literature (Table 1). For instance, construction material and reinforced structure were obtained from Guillard-Goncalves et al. [10], where an average vulnerability value was assigned to a study area located in Loures municipality. The scores obtained by Guillard-Goncalves et al. [10] were based on a questionnaire applied to a pool of 14 experts with field-based knowledge of landslides in the area north of Lisbon.

The weights assigned to each parameter of the physical vulnerability index (PVI) were based on a previously published work [18] applied to the municipality scale in Portugal, where the material construction has more importance, followed by the reinforced structure and number of floors. Because this index presents two more variables than Pereira et al. [18], it was necessary to adapt the weights, including the need for repairs to the structure and the need for repairs to the structure (NRS and NRF), which have the same weight as the construction period (PC) (0.1). The PVI was weighted according to Eq. (2):

$$PVI = 0.3CM + 0.2(RS + NF) + 0.1(PC + NRS + NRF) \quad (2)$$

Then, a cluster analysis was performed through the "Multivariate Clustering" tool available in ArcGIS Pro software, using the k means method to divide the buildings into four different groups, depending on the PVI parameters.

4 Results

4.1 Landslide Susceptibility and Exposure

Table 2 presents the predisposition factors and their weights, with the respective classes of each predisposition factor and each score, and Fig. 5 presents the spatial distribution of each predisposing factor.

Slope angle comprehends nine classes: <5; 5–10; 10–15; 15–20; 20–25; 25–30; 30–35; 35–40; >40, where classes with a higher slope angle present a higher score. Lithology has 15 classes: Volcano-sedimentary complex; Limestones (Jurassic); Sandstones (Jurassic); Marls and limestones (Jurassic); Limestones (Cretaceous); Marls and limestones; Sandstones (Cretaceous); 8—Volcanic complex of Lisbon; Detrital rocks (Paleogene); Other rocks (Miocene); Clayey rocks (Miocene); Detrital rocks (Pliocene); Quaternary deposits; Granite; Other acid magmatic rocks. TPI classes include V-shaped valleys; Moderate fit valleys, half slope drainage; U-shaped valleys; Flat areas, and slope base; Middle slope; Upper slope; and Top slope. In LMA, there are two classes of PGA, 2.4–3.2 and 3.2–4.0. Distance to faults' buffers is divided into five classes, with scores deriving according to the proximity. Slope curvature has three classes—convex, rectilinear, and concave—with a higher score for the concave areas, where a greater deposition of soil occurs.

The variable weights achieved by the Analytic Hierarchy Process, based on Saaty's scale of influence [23], were: slope, with a weight of 0.460,lithology, with a weight of 0.256; TPI, with a weight of 0.114; PGA, with a weight of 0.087; distance to faults, with a weight of 0.051; and curvature, with a weight of 0.032. Since the AHP susceptibility model obtained a consistency index of 0.06 and a consistency ratio of 0.05, which are <0.1, we can assume that our metric is reasonably consistent in the process of decision-making.

The quality of the susceptibility models was assessed through the computation of the ROC success curve and the AUC. The success curve consists of the comparison between the image of the susceptibility model and the identified slope movements used in the model. Figure 6 shows the ROC success rate curve of the rainfall-triggered landslide (RTL)) susceptibility model and the earthquake-triggered landslide (ETL) susceptibility model, according to the proportion of area correctly classified as susceptible (TPR) and the proportion of non-landslide area classified as susceptible (FPR). The ETL model hits with high quality the proportion of identified landslide inventory, with an AUC = 0.93, while the ETL model has a very satisfactory quality with an AUC of 0.82.

Figure 7 presents (A) the susceptibility map for landslides triggered by precipitation; and (B) the susceptibility map for landslides triggered by seismic events. The susceptibility is represented in 5 classes divided into quintiles—Very Low, Low, Moderate, High, and Very High.

Both landslide susceptibility models were used to filter buildings located in the most susceptible areas (90th percentile) to RTL and ETL (see Figs. 8s and 9a, respectively). Of the total number of buildings in the LMA (449,473), there are 10,201

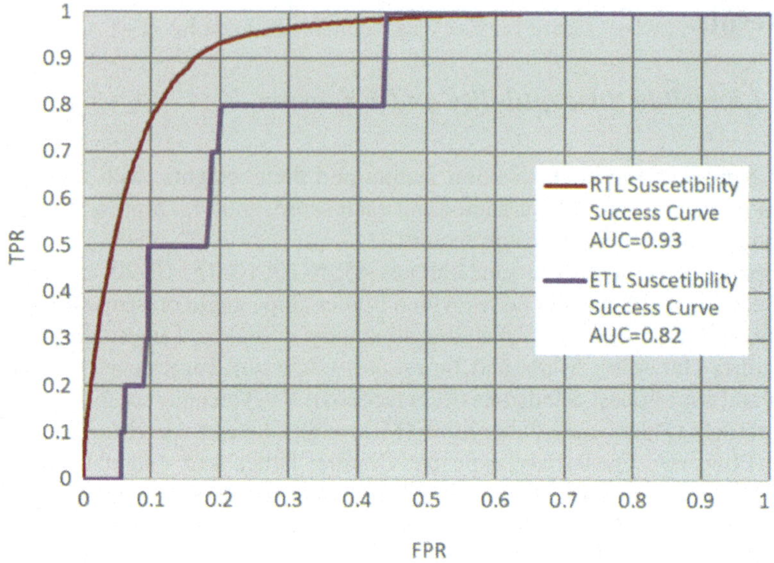

Fig. 6 ROC success rate curves of RTL and ETL susceptibility models

buildings in the areas most susceptible to RTL and 38,730 buildings in the areas most susceptible to ETL.

4.2 Physical Vulnerability of Buildings

Figure 8a shows the physical vulnerability of buildings (PVB) exposed to the 90th percentile of susceptibility to RTL, and Fig. 8b represents the average physical vulnerability per parish for a slip surface area of 1 m. Physical vulnerability is represented by five classes (Very Low, Low, Moderate, High, and Very High), divided into quintiles. The municipalities of Lisbon and Loures present the buildings with the highest physical vulnerability index in the case of RTL. Cascais, Oeiras, and Odivelas municipalities present the lowest average vulnerability values per parish.

Figure 9a shows the physical vulnerability of buildings exposed to the 90th percentile of susceptibility to ETL and Fig. 9b represents the average physical vulnerability per parish for a slip surface area of 10 m. Physical vulnerability is represented by five classes divided into quintiles. The municipalities of Lisbon present the buildings with the highest physical vulnerability index in the case of ETL. Cascais and Oeiras municipalities present the lowest average vulnerability values per parish.

The physical vulnerability to RTL ranges from 0.28 (minimum value and 1st quartile) to 0.71 (maximum value), records and an average value of 0.34 and the median corresponds to 0.32. The physical vulnerability index for ETL ranges between 0.585 and 0.89, records an average value of 0.65 and the median is 0.645. The PVB

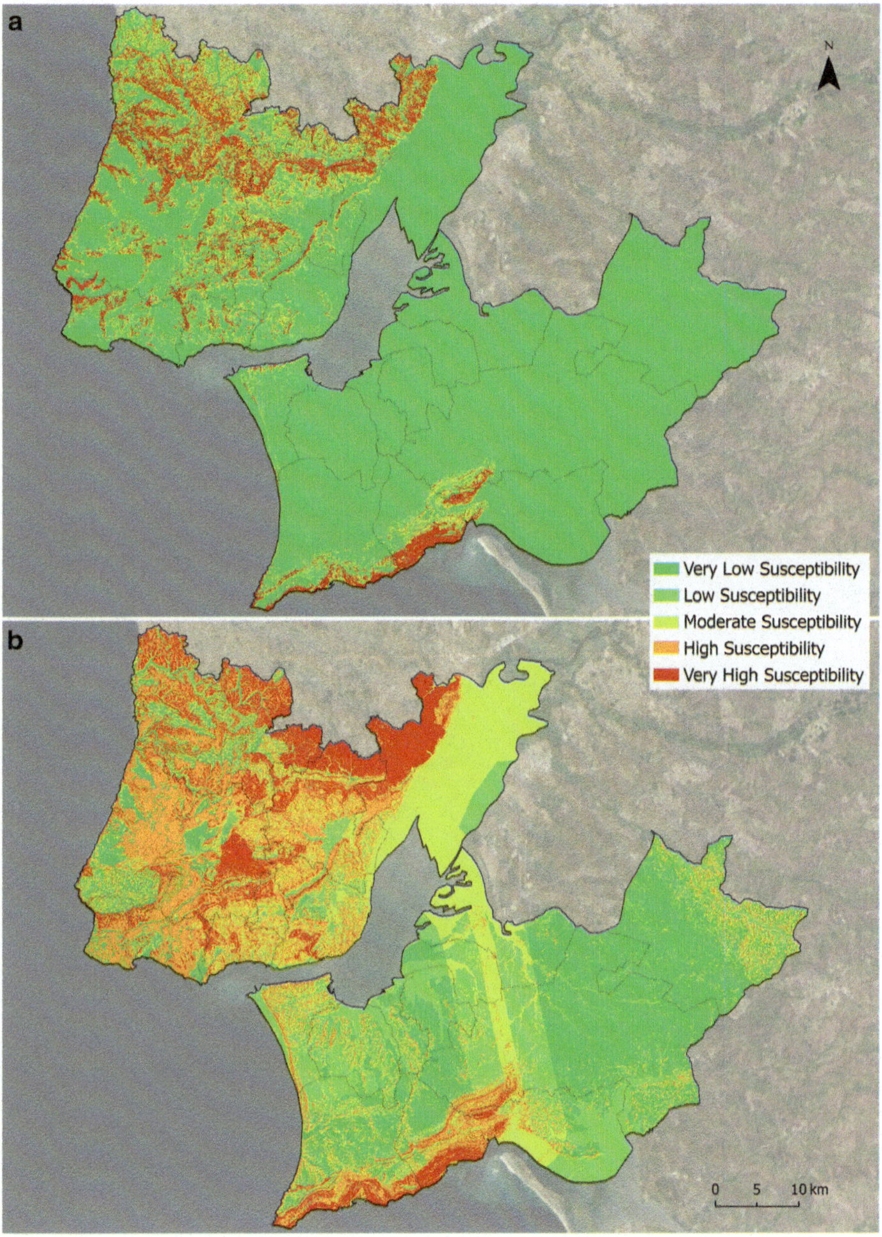

Fig. 7 **a** RTL susceptibility and **b** ETL susceptibility in the LMA

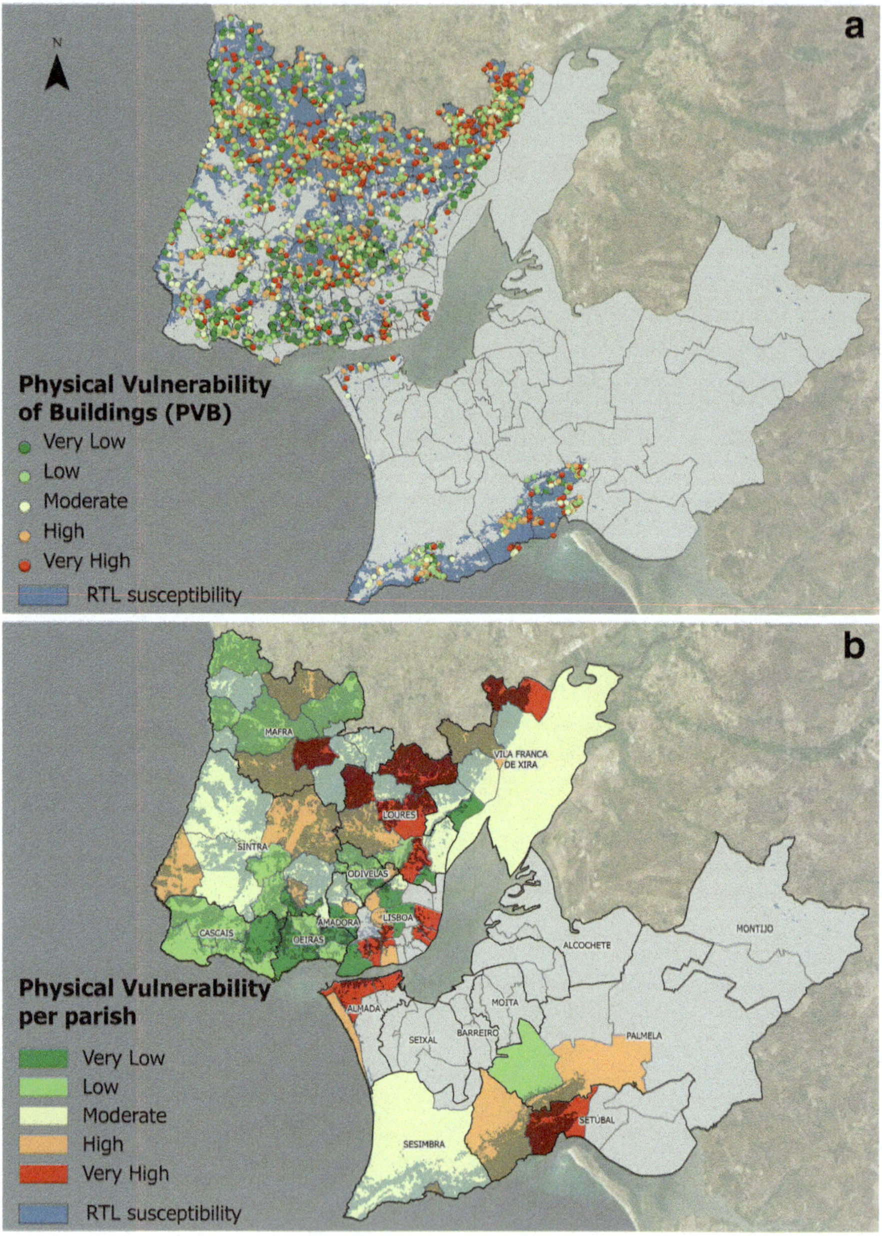

Fig. 8 Physical vulnerability of buildings (**a**) and the average physical vulnerability scores per parish (**b**) of buildings located in the 90th percentile of RTL susceptibility in the LMA (1 m slip surface scenario)

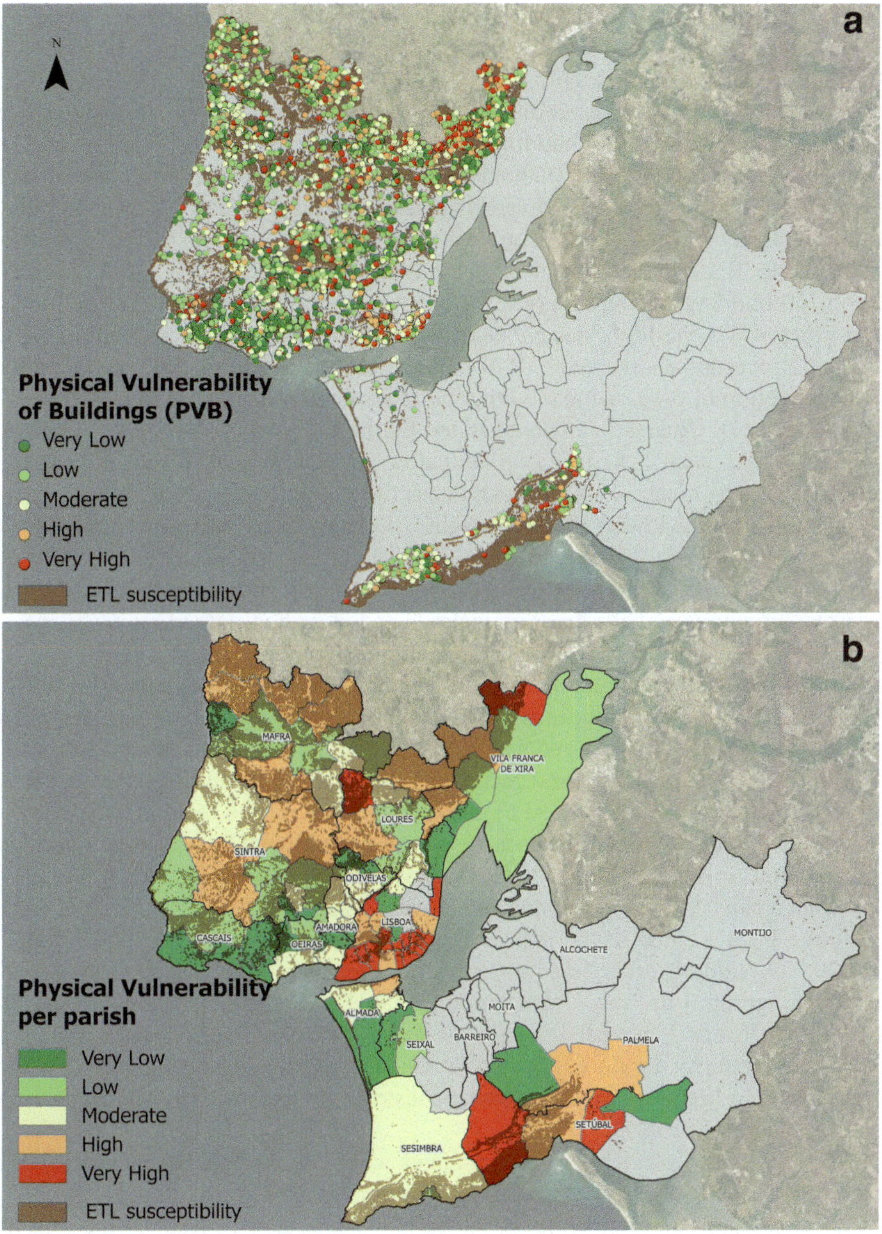

Fig. 9 Physical vulnerability of buildings (**a**) and the average physical vulnerability scores per parish (**b**) of buildings located in the 90th percentile of ETL susceptibility in the LMA (10 m slip surface scenario)

for RTL presents lower values than for ETL, which reflects the difference in the scores attributed to the classes of variables corresponding to the construction material and the reinforced structure, for both scenarios.

Figure 10 presents a spatial distribution of the cluster analysis, performed with the physical vulnerability of buildings exposed to the most susceptible areas of RTL (Fig. 10a) and ETL (Fig. 10b). In both landslide triggering scenarios, 4 clusters were identified. Figure 11 shows the physical vulnerability parameters standardized values for each cluster in the most susceptible areas to RTL (a) and ETL (b).

In the PVB located in the most susceptible areas of RTL (Fig. 10a), Group 1 has 2,760 buildings, group 2 has 1,397 buildings, group 3 has only 295 buildings, and group 4 has the highest number of buildings (5,749). Buildings belonging to Cluster 1 recorded an average PVI of 0.35, a number of floors of 0.5, and a period of construction of 0.43, while the other parameters recorded an average PVI below 0.35 (Fig. 11a). Cluster 1 is characterized by the presence of buildings one floor, constructed between 1961 and 1990 and after 1991, with more resistant structures and construction material and in a good state of maintenance.

Cluster 4 has the lowest PVI mean value (0.3) and period of construction (0.38), and the other parameters are below 0.35. This means that buildings included in Cluster 4 are mainly constructed with two or more floors, more recently, after 1991, with more resistant structures and construction material and in a good state of maintenance.

Cluster 2 recorded a mean PVI of 0.45, period of construction (0.65), with need to repairs ($\geq$0.6), and construction material and reinforced structure with a mean PVI below 0.4. This cluster includes more ancient buildings, with slab or masonry walls, with less resistant structures and need for repairs.

Cluster 3 presents the highest mean PVI (0.55), construction material and reinforced structure (0.6), and period of construction (0.65). The buildings included in this cluster are more vulnerable because they were built with less resistant structures and materials in a period before 1919 and between 1919–1945. Buildings belonging to cluster 3 can be found mostly in Loures, Mafra, and Vila Franca de Xira' parishes.

In the PVB located in the most susceptible areas of ETL (Fig. 10b), Group 1 has 20,239 buildings, this being the most representative group; group 2 has 4,477 buildings; group 3 has the fewest buildings (3,812); and group 4 contains 10,202 buildings. Cluster 1 recorded the lowest mean PVI (0.6), period of construction (0.45), and number of floors (0.3). We can find more resistant buildings in this cluster, with a higher number of floors, constructed more recently (between 1961 and 1990 and after 1991).

Cluster 4 has slightly higher mean values when compared with cluster 1, but it records a higher PVI associated with the number of floors (1 or 2 floors).

Clusters 2 and 3 are very similar in all building parameters, except for the need for repairs, which has higher PVI mean values in cluster 2. In cluster 2, it is also possible to observe the highest mean PVI of all clusters. Buildings belonging to cluster 2 can be found mostly in Lisbon, Sintra, and Vila Franca de Xira parishes.

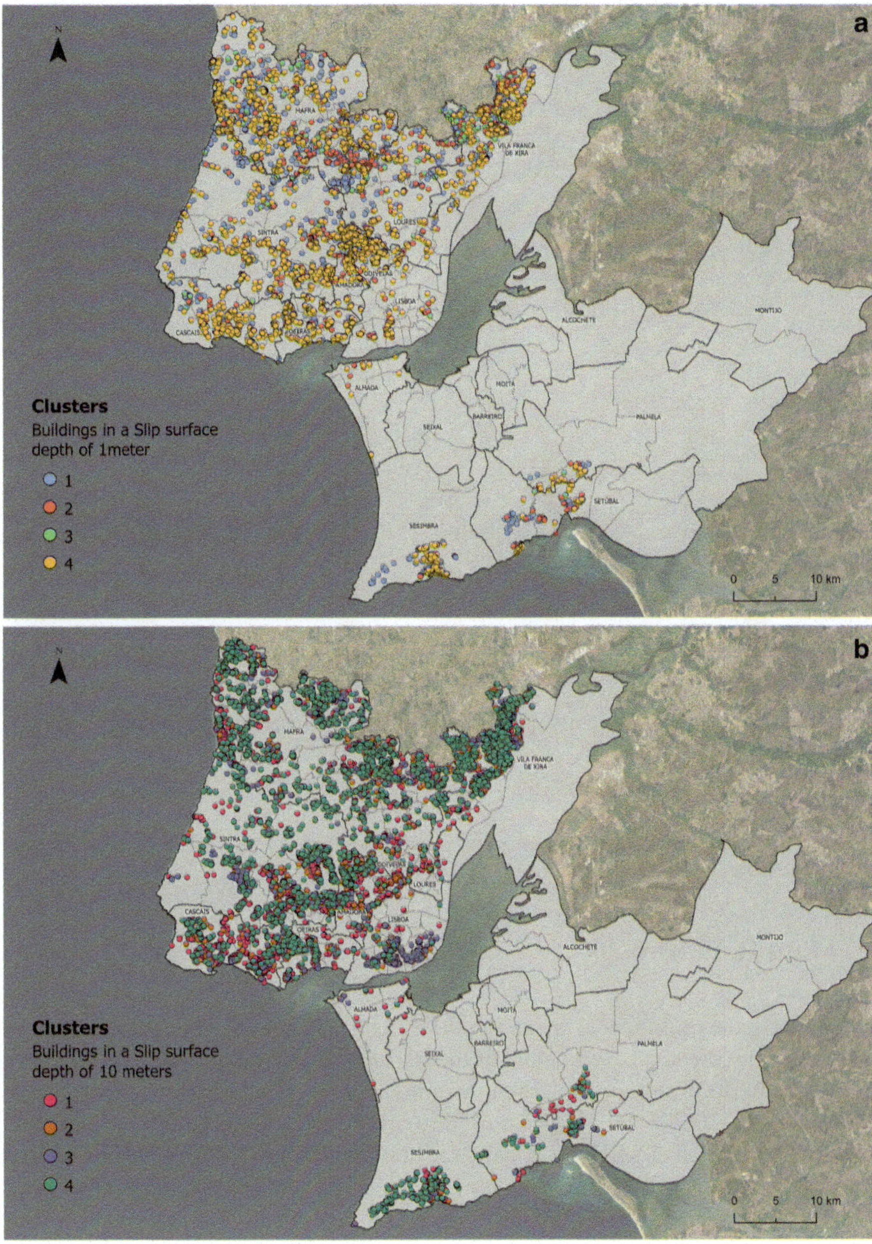

Fig. 10 Clusters of PVB located in the most susceptible areas to RTL (**a**) and ETL (**b**) scenarios

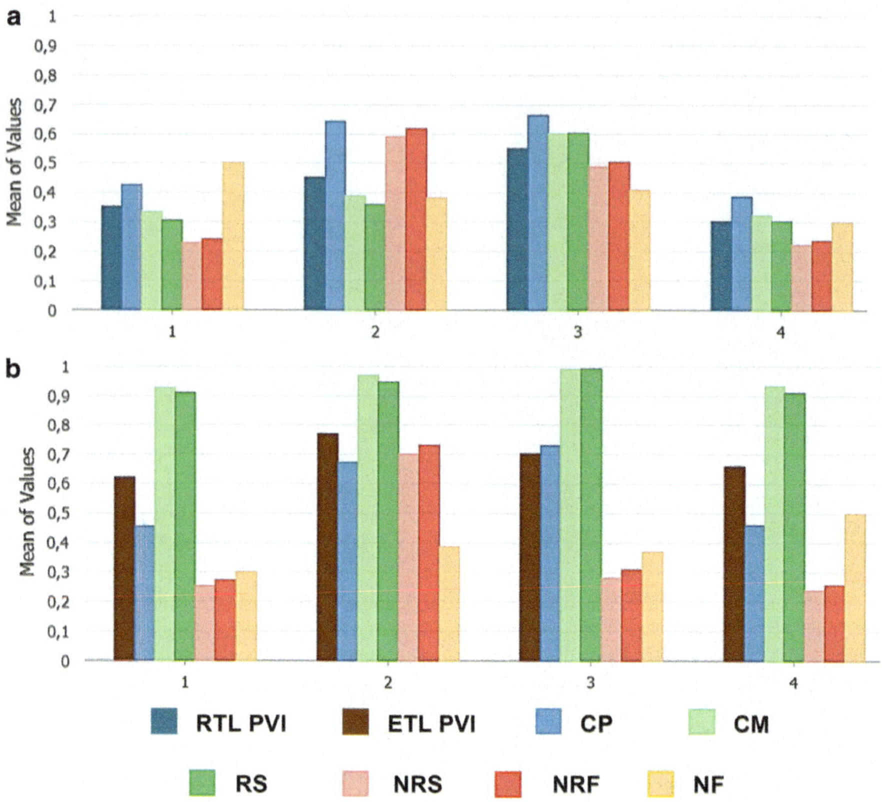

Fig. 11 Physical vulnerability parameters and corresponding mean values of PVI for each cluster in the most susceptible areas to RTL (**a**) and ETL (**b**) scenarios

5 Final Remarks

In this work, the buildings exposed to two landslide triggering scenarios were identified, and an indicator-based model of the PVB was carried out to find the buildings most vulnerable to this natural phenomenon. The two scenarios correspond to the analysis of the 10 201 buildings exposed to a very high susceptibility to landslides triggered by rainfall (RTL) (Fig. 8); and the analysis of the 38,730 buildings exposed to a very high susceptibility to slope movements triggered by seismic events (Fig. 9).

A landslide susceptibility model was produced for each triggering scenario, although it is necessary to highlight the limitations associated with each susceptibility model validation. The RTL susceptibility model was validated through photo-interpretation inventories (4,047 landslides) and some field validation. The ETL susceptibility model, on the other hand, was validated through a rather small historical inventory (10 landslides). The location of the buildings was not very accurate, as well.

In this way, it would be important in the future to attest to the quality of models with more recent events. The susceptibility to ETL was modeled through the Analytic Hierarchy Process (multicriteria analysis method), which implies a subjective hierarchy, based on expert opinion and previous studies; lacking a ground validation of the quality of susceptibility and a validation of the model through occurrences of landslides triggered by future seismic events.

The very high susceptible areas to RTL are more restricted to the most hazardous slopes in the LMA (339.54 km^2) located in the hills in the north of Lisbon and in the Arrábida chain, and the very high susceptible areas to ETL (309.91 km^2) are around the same main locations but with larger features. This is the first time that these approaches have been evaluated jointly for the entire LMA.

The analysis of buildings exposed to slope movements in a scenario where the building is located in the slip surface area, for the two scenarios, was only carried out for areas with very high susceptibility; this is limiting the analysis to the 90th percentile of susceptibility. This does not mean that there cannot be damage to buildings belonging to areas with high susceptibility, for example, areas with an 80th percentile of susceptibility; however, in this work, it was assumed that we only worked with 10% of the most susceptible areas in the entire LMA, for both scenarios (RTL and ETL). Sliding failure of foundations, depth of the landslide, and impact direction affect the building damage. For instance, longitudinal impact leads to a larger contact area and a small resistance moment, which can generate severe damage to the buildings, although these scenarios were not considered. Future studies should consider a wide range of worst scenarios in landslide-building interaction studies.

The physical vulnerability index (PVI) was based on bibliographic support [10, 18]. Uncertainties can be quantified from the input parameters to the vulnerability estimates and the weights are often based on expert judgment, which seems subjective. However, for the LMA, there is not a good inventory of damages generated by landslides and the characteristics of the respective buildings affected. This makes it difficult to legitimize the scores and weights assigned to classes and variables, respectively and to apply a quantitative vulnerability model developed by statistical analysis or post-events data.

The variables used to perform the physical vulnerability index (PVI) were taken from the Building Georeferencing Base (BGE) of the 2011 census date. This information is out of date since there is already a 2021 census; however, information and data about the characteristics of the buildings were not made available at the time of this work, and there may be missing information regarding new buildings built in the last ten years.

In the LMA, it was found that the most problematic PVB areas, for both scenarios (RTL and ETL), correspond to the parish of Lisbon, presenting an average of buildings with higher physical vulnerability. There is a big difference in the variation of the PVB index values for a slip surface area of 1 m for an RTL (ranges between 0.28 and 0.71), and for a slip surface area of 10 m for an ETL (ranges between 0.585 and 0.89). This reflects the different scores attributed to the variables of construction material and reinforced structure in the two scenarios [10, 18], which correspond to the variables with greater weight in the PVI.

This study is carried out on a building scale for the entire territory of LMA, i.e., a large study area with a small scale. However, this work presents a lot of potential for analysis and future complements; for example, in a larger scale study, at the level of street/block analysis, through the identification of multi-risk hotspot areas and the identification of safe or evacuation areas for the resident population. Through the hotspots, it would be possible to identify priority risk prevention areas and implement local strategies to reduce the physical vulnerability of buildings and their exposure to landslides triggered by future rainfall and earthquake events, with detailed studies and detailed technical protection interventions (on-site intervention). This more in-depth study can also be carried out at the level of the municipality (making the hierarchy of parishes according to the average value of PVI) and at the level of the parish.

Acknowledgements The project 'MIT-RSC—Multi-risk Interactions Towards Resilient and Sustainable Cities' (MIT-EXPL/CS/0018/2019) leading to this work is co-financed by the ERDF—European Regional Development Fund through the Operational Program for Competitiveness and Internationalisation—COMPETE 2020, the North Portugal Regional Operational Program—NORTE 2020 and by the Portuguese Foundation for Science and Technology—GCT under the MIT Portugal Program at the 2019 PT call for Exploratory Proposals in 'Sustainable Cities'. Pedro P. Santos is financed through FCT I.P., under the contract CEECIND/00268/2017.

References

1. Alves C (2018) Padrões de vulnerabilidade estrutural em estradas e edifícios associados à ocorrência de movimentos de vertente [Dissertação de Mestrado]. Instituto de Geografia e Ordenamento do Território
2. Alves C, Oliveira S (2017) Vulnerabilidade física das estradas a deslizamentos. In: Gomes A, Teixeira S, Soares L (eds) Atas do VIII Congresso Nacional de Geomorfologia. Associação Portuguesa de Geomorfólogos, pp 63–66
3. Araújo JR, Ramos AM, Soares PMM, Melo R, Oliveira SC, Trigo RM (2022) Impact of extreme rainfall events on landslide activity in Portugal under climate change scenarios. Landslides. https://doi.org/10.1007/s10346-022-01895-7
4. Bernardo V, Sousa R, Candeias P, Costa A, Campos Costa A (2021) Historic appraisal review and geometric characterization of old masonry buildings in lisbon for seismic risk assessment. Int J Archit Herit, 1–21. https://doi.org/10.1080/15583058.2021.1918287
5. Cabral J (1995) Neotectónica em Portugal continental. Memórias Dos Serviços Geológicos de Portugal, 31
6. Cabral J, Moniz C, Ribeiro P, Terrinha P, Matias L (2003) Analysis of seismic reflection data as a tool for the seismotectonic assessment of a low activity intraplate basin—the Lower Tagus Valley (Portugal). J Seismolog 7(4):431–447. https://doi.org/10.1023/B:JOSE.0000005722.23106.8d
7. Daveau S (2004) A Estremadura. In: Feio M, Daveau S (eds) O relevo de Portugal: grandes unidades regionais, vol 2, pp 61–74. Associação Portuguesa de Geomorfólogos
8. Gordo C, Zêzere JL, Marques R (2019) Landslide susceptibility assessment at the basin scale for rainfall- and earthquake-triggered shallow slides. Geosciences 9(6). https://doi.org/10.3390/geosciences9060268
9. Guillard C, Zezere J (2012) Landslide susceptibility assessment and validation in the framework of municipal planning in Portugal: the case of Loures municipality. Environ Manage 50(4):721–735. https://doi.org/10.1007/s00267-012-9921-7

10. Guillard-Goncalves C, Zêzere JL, Pereira S, Garcia RAC (2016) Assessment of physical vulnerability of buildings and analysis of landslide risk at the municipal scale: application to the Loures municipality, Portugal. Nat Hazard 16(2):311–331. https://doi.org/10.5194/NHESS-16-311-2016
11. Kullberg JC, Rocha RB, Soares AF, Rey J, Terrinha P, Azerêdo AC, Callapez P, Duarte LV, Kullberg MC, Martins L, Miranda R, Alves C, Mata J, Madeira J, Mateus O, Moreira M, Nogueira CR (2013) A Bacia Lusitaniana: Estratigrafia, Paleogeografia e Tectónica. In: Dias R, Araújo A, Terrinha P, Kullberg JC (eds) Geologia de Portugal, vol II, 1st ed. Escolar Editora, pp 195–350
12. Malheiros J, Zêzere J, Ludovici A, Pereira S, Oliveira S, Malheiros M (2018) Um século de respostas habitacionais públicas a catástrofes—Experiências passadas e reflexões para o future, pp 365–405
13. Martins A (2004) As bacias sedimentares do Baixo Tejo e do Sado. In: Feio M, Daveau S (eds) O relevo de Portugal: grandes unidades regionais, vol 2, pp 49–60. Associação Portuguesa de Geomorfólogos
14. Oliveira S (2012) Incidência espacial e temporal da instabilidade geomorfológica na bacia do Rio Grande da Pipa (Arruda dos Vinhos) [Dissertação de Doutoramento]. Instiuto de Geografia e Oedenamento do Território
15. Oliveira S, Zêzere JL, Garcia RAC, Pereira S (2016) Padrão de deformação de movimentos de vertente em áreas periurbanas associados a eventos de instabilidade. In: International conference on urban risks: Atas ICUR2016, pp 363–370
16. Peláez Montilla JA, López Casado C (2002) Seismic hazard estimate at the Iberian Peninsula. Pure Appl Geophys 159(11):2699–2713. https://doi.org/10.1007/s00024-002-8754-3
17. Pereira DI, Pereira PJS, Santos LJC, Silva JM (2014) Unidades Geomorfológicas de Portugal Continental. Revista Brasileira de Geomorfologia 15(4):567–584. https://doi.org/10.20502/rbg.v15i4.549
18. Pereira S, Santos PP, Zêzere JL, Tavares AO, Garcia RAC, Oliveira SC (2020) A landslide risk index for municipal land use planning in Portugal. Sci Total Environ 735:139463. https://doi.org/10.1016/j.scitotenv.2020.139463
19. Pina C, Alvarenga M, Pereira L (2019) O Ordenamento do Território na Resposta às Alterações Climáticas Contributo para os PDM
20. Ramos C, Ramos-Pereira A (2020) Landscapes of Portugal: paleogeographic evolution, tectonics and geomorphology. In Vieira G, Zêzere JL, Mora C (eds) Landscapes and landforms of Portugal. Springer International Publishing, pp 3–31. https://doi.org/10.1007/978-3-319-03641-0_1
21. Ramos-Pereira A (2003) Geografia Física e Ambiente—Diversidade do Meio Físico e Recursos Naturais. Atlas Da Área Metropolitana de Lisboa. Área Metropolitana de Lisboa, Lisboa, pp 47–65
22. Roque A (2007) Tectonostratigrafia do Cenozóico das margens continentais Sul e Sudoeste portuguesas: um modelo de correlação sismostratigráfica [Tese de doutoramento, Universidade de Lisboa]. http://hdl.handle.net/10451/1521
23. Saaty T (2008) Decision making with the analytic hierarchy process. Int J Serv Sci 1:83–98. https://doi.org/10.1504/IJSSCI.2008.017590
24. Santos M, Fonseca A, Fragoso M, Santos JA (2019) Evolução recente e futura de índices de extremos de precipitação em Portugal continental. In: Pereira A, Leal M, Bergonse R, Trindade J, Reis E (eds) Água e Território: um tributo a Catarina Ramos, Issue July. Universidade de Lisboa, Centro de Estudos Geográficos, pp 279–294. https://doi.org/10.33787/CEG20190005
25. Senos ML, Carrilho F (2003) Sismicidade de Portugal continental. Física de La Tierra 15:93–110
26. Silva M, Pereira S (2014) Assessment of physical vulnerability and potential losses of buildings due to shallow slides. Nat Hazards 72(2):1029–1050. https://doi.org/10.1007/s11069-014-1052-4
27. Sterlacchini S, Akbas S, Blahůt J, Mavrouli O, Garcia C, Quan Luna B, Corominas J (2014) Methods for the characterization of the vulnerability of elements at risk. In: Advances in natural

and technological hazards research, vol 34, pp 233–273. https://doi.org/10.1007/978-94-007-6769-0_8
28. van Westen C, Kingma N (2011) Session 5: vulnerability assessment. In: van Westen CJ, Alkema D, Damen MCJ, Kerle N, Kingma NC (eds) Multi-hazard risk assessment: distance education course/Guide book. United Nations University–ITC School on Disaster Geoinformation Management, pp 5; 1–5; 33
29. Vaz T (2010) Contribuição para o estudo dos movimentos de vertente desencadeados por eventos sísmicos em Portugal Continental [Msc Dissertation, Universidade de Lisboa]. https://repositorio.ul.pt/handle/10451/3873
30. Vaz T, Zêzere JL (2016) Landslides and other geomorphologic and hydrologic effects induced by earthquakes in Portugal. Nat Haz 81(1):71–98. https://doi.org/10.1007/S11069-015-2071-5/FIGURES/11
31. Vaz T, Zêzere JL (2020) The urban geomorphological landscape of Lisbon. In: Vieira G, Zêzere JL, Mora C (eds) Landscapes and landforms of Portugal. World geomorphological landscapes. Springer, p. 295–303. https://doi.org/10.1007/978-3-319-03641-0_23
32. Westen CJ, Castellanos E, Kuriakose S (2008) Spatial data for landslide susceptibility, hazard, and vulnerability assessment: an overview. Eng Geol 102:112–131. https://doi.org/10.1016/j.enggeo.2008.03.010
33. Xofi M (2021) Evaluation of exposure and vulnerability indicators for assessing multi-hazard risk in urban areas. Am J Orthod Dentofac Orthop, no 2. https://doi.org/10.1016/j.ajodo.2021.06.008
34. Yin KL, Yan TZ (1988) Statistical prediction models for instability of metamorphosed rocks. Int Symp Lands 5:1269–1272
35. Zêzere JL (2002) Landslide susceptibility assessment considering landslide typology. A case study in the area north of Lisbon (Portugal). Nat Haz Earth Syst Sci 2(1–2):73–82. https://doi.org/10.5194/NHESS-2-73-2002
36. Zêzere JL (2020) Geomorphological hazards. In: Vieira G, Zêzere JL, Mora C (eds) Landscapes and landforms of Portugal. Springer International Publishing, pp 47–62. https://doi.org/10.1007/978-3-319-03641-0_3
37. Zêzere JL, Pereira S, Tavares AO, Bateira C, Trigo RM, Quaresma I, Santos PP, Santos M, Verde J (2014) DISASTER: a GIS database on hydro-geomorphologic disasters in Portugal. Nat Haz 72(2):503–532. https://doi.org/10.1007/S11069-013-1018-Y
38. Zêzere JL, Trigo RM, Fragoso M, Oliveira SC, Garcia RAC (2008) Rainfall-triggered landslides in the Lisbon region over 2006 and relationships with the North Atlantic oscillation. Nat Haz Earth Syst Sci 8(3):483–499. https://doi.org/10.5194/NHESS-8-483-2008
39. Sintra Notícias. (2018 March 15) Morreu uma das vítimas na derrocada de fa-lésia na praia da Ursa. Sintra Notícias. https://sintranoticias.pt/2018/03/15/ultima-hora-derrocada-falesia-na-praia-da-ursa-sintra/

Chapter 8
Multi-hazard Susceptibility Assessment for Land Use Planning in the Lisbon Metropolitan Area

José Luís Zêzere, Ricardo A. C. Garcia, Raquel Melo, Sérgio Cruz Oliveira, Susana Pereira, Eusébio Reis, Ângela Santos, and Pedro Pinto Santos

Abstract The Lisbon Metropolitan Area is a risk hotspot in Portugal due to excessive exposure to natural and environmental hazards. In this work, a multi-hazard susceptibility assessment is performed for the 118 parishes that constitute the study area, considering the spatial incidence of seven hazardous processes: earthquakes, tsunami, beach erosion and coastal flooding, coastal erosion and cliff retreat, landslides, floods, and forest fires. The relative importance of hazardous processes was established through the Analytic Hierarchy Process (AHP), based on the frequency-magnitude relationship of each process and its damage capacity. All the parishes exhibiting very high multi-hazard susceptibility have high earthquake susceptibility and most are located in riverine or coastal zones, thus subjected to floods and/or coastal erosion (affecting beaches and/or cliffs).

J. L. Zêzere (✉) · R. A. C. Garcia · S. C. Oliveira · S. Pereira · E. Reis · Â. Santos · P. P. Santos
Centre for Geographical Studies, Institute of Geography and Spatial Planning (IGOT), LA TERRA, University of Lisbon, Lisbon, Portugal
e-mail: zezere@igot.ulisboa.pt

S. C. Oliveira
e-mail: cruzdeoliveira@campus.ul.pt

E. Reis
e-mail: eusebioreis@edu.ulisboa.pt

Â. Santos
e-mail: angela.santos@campus.ul.pt

P. P. Santos
e-mail: pmpsantos@campus.ul.pt

R. Melo
Institute of Earth Sciences, School of Science and Technology, University of Évora, Évora, Portugal
e-mail: raquel.melo@uevora.pt

S. Pereira
Faculty of Arts and Humanities, Geography Department, University of Porto, Porto, Portugal
e-mail: sspereira@letras.up.pt

T. M. Ferreira (ed.), *Multi-risk Interactions Towards Resilient and Sustainable Cities*, Advances in Sustainability Science and Technology,
https://doi.org/10.1007/978-981-99-0745-8_8

Keywords Multi-hazard · Susceptibility · Mapping · Analytic hierarchy process · Lisbon metropolitan area

1 Introduction

The Lisbon Metropolitan Area (LMA) covers 3,015 km^2 (3.3% of mainland Portugal) and is distributed by 18 municipalities and 118 parishes. In 2021, the resident population amounted to 2,821,876 inhabitants (1.7% more than in 2011), about a quarter of the Portuguese population. In economic terms, the LMA concentrates nearly 25% of the active population, 30% of national companies, and 33% of employment, and contributes to more than 36% of the national GDP [1].

The LMA has a remarkable geomorphological diversity and encompasses an Atlantic coastline of 150 km and a riverfront of around 200 km. It includes one large estuary (Tagus) and part of another (Sado), and five protected areas, integrated into the Natura 2000 Network. The elevation ranges from 0 to 528 m a.s.l (Fig. 1).

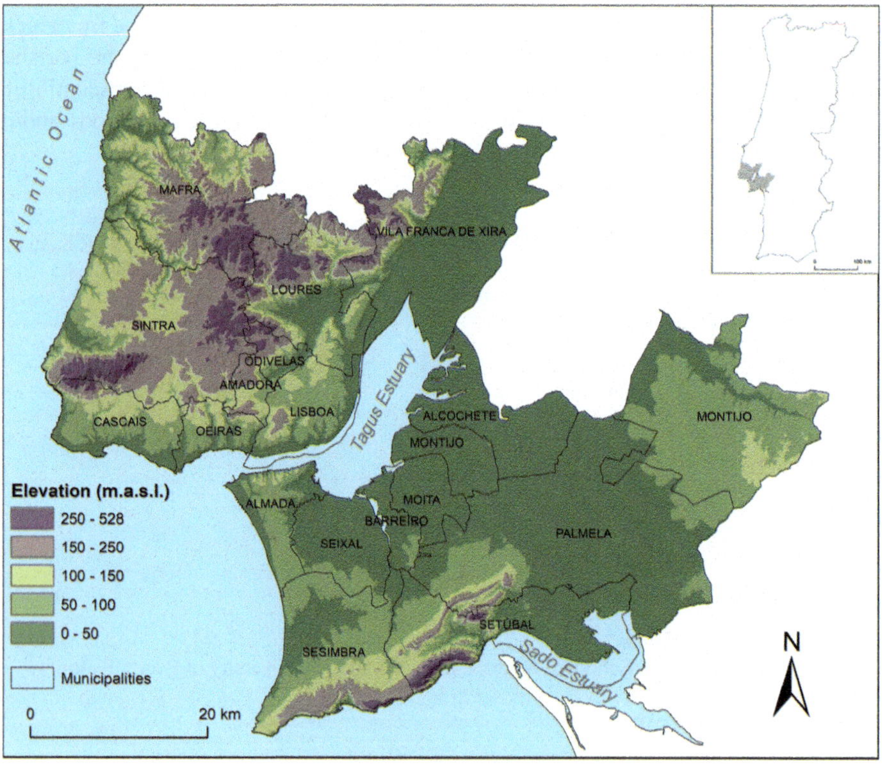

Fig. 1 Location and elevation of the Lisbon Metropolitan Area

The LMA is a major risk hotspot in Portugal due to excessive exposure to natural hazards like earthquakes, floods, and coastal erosion [24]. This area suffered the effects of the two greatest natural disasters that occurred in mainland Portugal in the last 300 years: the Lisbon earthquake and tsunami on November 1, 1755 [3], that caused more than 12,000 deaths, and the flash flood during November 25–26, 1967 [21], that caused more than 700 deaths.

The existing hazards have been causing a significant impact in the study area and need to be accounted for by urban planning and civil protection stakeholders. Therefore, this chapter aims to assess the combined susceptibility of a set of natural and environmental hazards relevant to regional and urban planning in the LMA. The assessment is performed for the 118 parishes that constitute the study area, and it will contribute to improving preventive land use planning by bridging the gap between scientific research and land use planners.

2 Data and Methods

The susceptibility was assessed and mapped for a set of hazardous physical processes with a relevant territorial incidence in the LMA: earthquakes, tsunamis, beach erosion and coastal flooding, coastal erosion and cliff retreat, landslides, floods, and forest fires.

In the first stage, the susceptibility assessment was carried out separately for each type of hazardous process, using recognized but heterogeneous technical-scientific methods adjusted to the regional scale and available data. Details on the methods used are provided in the following subsections.

The susceptibility for each hazardous process was re-scaled for the parish level into a scale ranging from 1 to 5, following a set of rules that are described in Sect. 4.

In the second stage, the Analytic Hierarchy Process (AHP) [15, 16] was used to weigh the relative importance of the hazardous processes considered in the study area. The AHP follows three successive steps [15, 16]: (i) decomposition of the problem into criteria; (ii) parity comparative analysis between the criteria by means of a numerical scale; and (iii) synthesis of priorities by means of eigenvector calculation or least square analysis.

The results provided by the AHP were used to weigh the re-scaled susceptibility scores of each individual hazardous process, which were later added together to obtain the multi-hazard susceptibility at the parish level for the complete LMA. These results are presented with five qualitative classes (very high, high, medium, low, and very low) that were established based on the standard deviation method.

2.1 Earthquakes

The seismic hazard in the LMA is generally high due not only to the proximity of active submarine tectonic structures bordering the Portuguese territory to the SW and S, which have the potential to generate the maximum regional earthquakes [5], but also to the fault (or fault zone) of the lower Tagus valley [4].

The Regulation of Safety and Actions for Buildings and Bridge Structures, approved by Decree-law n°235/83 of 31 May, established a seismic risk zoning in mainland Portugal and defined the safety coefficients to be applied in building and bridge constructions in each of the four zones identified. The LMA fits in zone A, which is the highest risk zone.

The seismic susceptibility was defined from the crossing of the isoseismal map of maximum seismic intensity (Modified Mercalli Scale; Source: IPMA) with the Peak Ground Acceleration (PGA) for a return period of 475 years, provided by Peláez and López-Casado [12]. Additionally, the site effects that are likely to produce the amplification of seismic susceptibility were considered, based on two sources: (i) the distribution of unconsolidated sedimentary geological formations, represented in the Geological Map of Portugal at scale 1:500,000 (Source: LNEG); and (ii) 100 m-buffer zones along the active or probably active faults extracted from the Neotectonic Map of Portugal [2].

2.2 Tsunami

The Portuguese coastal zone is exposed to the tsunami hazard associated with earthquake events with epicenter at the sea floor. Destructive earthquake-triggered tsunamis are more probable to be generated in three major regional seismic zones: (i) the Gorringe Bank, SW of Portugal; (ii) active tectonic structures of N-S direction, along the continental margin between Setúbal and Cape S. Vicente; (iii) and the eastern terminus of the Azores-Gibraltar fault, south of the Algarve.

The tsunami susceptibility was assessed with a deterministic approach, using formulas published by Okada [10]. The event of November 1, 1755, is assumed to be the worst-case scenario, with a possible generation source located in the Gorringe Bank [17] and assuming the following earthquake parameters: length = 200 km, width = 80 km, azimuth = 60°, dip angle = 40°, displacement angle = 90°, depth = 8 km, and vertical displacement = 12.1 m.

The tsunami propagation was modeled with the TUNAMI code (Tohoku University, Japan), following the methodology of Imamura [7], in which the equations of the Shallow Water Model are implemented and discretized with the leap-frog finite difference scheme. The model also considers seabed friction, according to the Manning coefficient $n = 0.025$.

The numerical tsunami model was applied to 5 computational regions; each region is embedded in the previous one, with a smaller area and smaller cell size. Bathymetry data at various scales (GEBCO Digital Atlas, 2003, Nautical charts of the Hydrographic Institute, 2005, 2009, 2012) were used in the construction of each computational region. Computational region 5 is based on topographic maps at scale 1:25,000, which were used to build a detailed digital terrain model (DTM) with a cell size of 25 m. This DTM was used to assess the tsunami flooding along the complete coastal and estuarine zones of the LMA.

The water height associated with the tsunami was classified into 2 classes: 0.1–2 m and more than 2 m, corresponding to tsunami flood susceptibility high and very high, respectively.

2.3 Beach Erosion and Coastal Flooding

The characterization and delimitation of coastal zones experiencing beach erosion and coastal flooding are based on the work promoted by the Portuguese Environment Agency (APA) that supported two Coastal Zone Programmes (Alcobaça-Cabo Espichel and Espichel-Odeceixe).

In the original work, the susceptibility of beach erosion and coastal flooding was assessed for two time periods: 2050 and 2100 [18, 19]. This assessment included the effects of climate change, in particular the rise of mean sea level, together with the future evolution of the coastline, derived from the long-term trend based on the evolution observed in the last 50 years. In addition, the potential for an "instantaneous" retreat of the beach profile (and of the coastline) when acted upon by an extreme storm, with different return periods, was also considered. Therefore, the susceptibility zones combine the cumulative action of the following processes: the long-term shoreline evolution associated with sediment deficit, the retreat due to the ongoing sea-level rise, and the retreat due to storm surge and the wave runup in storm conditions [18, 19]. The retreat zone projected for 2050 was considered of very high susceptibility, while the additional retreat zone for 2100 was considered of high susceptibility.

In the original work of Silva and co-authors [18, 19], the coastal retreat was not classified where beaches are bordered by a cliff or any anthropogenic structure, regardless of the temporal scenario. In these cases, a flood elevation of 8 m above the mean sea level was used as a reference, taking into consideration the mean sea-level variation, the maximum astronomical high tide, the maximum recorded storm surge, and the wave runup. This value is in line with those obtained for the coastal zone of Sintra and Cascais municipalities [8, 20].

2.4 Coastal Erosion and Cliff Retreat

As in the case of beach erosion and coastal flooding, the characterization and delimitation of coastal erosion and cliff retreat are based on research promoted by the Portuguese Environment Agency (APA), which supported the Coastal Zone Programmes Alcobaça-Cabo Espichel and Espichel-Odeceixe [9, 13, 14].

The original work was supported by a systematic inventory of landslides affecting cliffs using photo-interpretation techniques applied to a series of vertical aerial photographs obtained from the late 1940s to 2010 [13]. Moreover, two susceptibility zones adjacent to the cliff crest were defined in the original study, considering the maximum size of landslides affecting each homogeneous cliff sector, together with the geological and geomorphological characteristics of the cliffs [9, 13, 14]. These susceptibility zones were named "I—risk zone" and "II—additional zone", and their extension varies depending on the cliff height (from 0.5 to 2 times the cliff height). In this work, the "I—risk zone" is considered as very highly susceptible to cliff retreat, whereas the "II—additional zone" is considered as highly susceptible to cliff retreat.

2.5 Landslides

The landslide susceptibility was assessed for a larger area bringing together the LMA with three other NUTS III (Oeste, Lezíria do Tejo, and Médio Tejo). The landslide susceptibility modelling was based on the Information Value statistical method [22, 23], using seven landslide predisposing factors: slope, aspect, curvature, topographic position index, topographic humidity index, lithology, and land use. These factors were crossed with a landslide inventory containing 4047 rainfall-triggered landslides. The landslide susceptibility model was validated using the success rate curve and computing the Area Under the Curve (AUC = 0.92).

The landslide susceptibility map contains 5 classes (very high, high, moderate, low, and very low) that were defined based on the slope breaks of the success rate curve of the predictive model, and the corresponding percentage of validated landslide-affected areas is 50, 20, 20, 5, and 5%, respectively.

2.6 Floods

The assessment of flood susceptibility took into consideration the available documentation from previous works, namely (i) the flood areas identified by the Flood Risk Management Plans that were promoted by the Portuguese Environment Agency (APA); (ii) the areas threatened by floods identified by the Regional Framework of the LMA National Ecological Reserve; (iii) the delimitation of the extreme flood that occurred along the Tagus River in February 1979 and the delimitation of the centennial flood in the Sado estuary area, produced by the National Civil Engineering Laboratory (LNEC); (iv) the flood extent in the Tagus estuary proposed by Guerreiro et al. [6], based on hydrodynamic modeling; and (v) a set of areas threatened by floods, outlined at the municipal level within the delimitation of the National Ecological Reserve.

The 100-year return period was considered as the flood reference. Riverine floods were distinguished from flash floods, the former occurring along the main river in the region (Tagus River), the latter occurring in small and medium-sized hydrographic basins. Estuarine floods were also considered separately, namely in the Tagus and Sado rivers.

Estuarine floods are generally triggered by storm surges together with the effect of the tide, whereas riverine floods are typically triggered by rainy periods lasting over several weeks, drastically diminishing the regulating effect of dams, which may potentiate flood peaks with the respective discharges. On the contrary, flash floods occur in hydrographic basins with a short concentration time and in response to episodes of very intense rainfall concentrated in a few hours.

2.7 Forest Fires

The forest fire susceptibility was assessed using a Bayesian-based bivariate statistical method: the Likelihood Ratio (LR), applied to mainland Portugal [11]. The predictive model uses slope, elevation, and land use/land cover as conditioning factors, whereas the dependent variable corresponds to the burnt areas registered in the period 1975–2018. The final expression of forest fire propensity was computed through the product of the LR scores by the simple probability of fire occurrence. The latter was obtained for each 25 m cell size as the ratio between the number of times burned and the total number of years within the burnt area's database (44 years). The forest fire predictive model was validated by computing the success rate curve and estimating the Area Under the Curve (AUC = 0.83). The forest fire susceptibility map contains 5 classes (very high, high, moderate, low, and very low) that were defined based on the slope breaks of the success rate curve of the predictive model [11]. The corresponding percentage of validated burnt areas at the national scale is as follows: 50, 40.6, 6.6, 2.8, and 0%.

3 Single-Hazard Susceptibility Assessment

3.1 Earthquakes

The earthquake susceptibility is shown in Fig. 2 and encompasses 3 classes. The very high susceptibility class covers 20.6% of the LMA and corresponds to the areas with, simultaneously, the seismic intensity of degree IX or X, PGA ranging from 3.2 to 4.0 m/s^2, and the existence of unconsolidated Quaternary deposits and/or active faults. The high susceptibility class covers 38% of the LMA and corresponds to the following

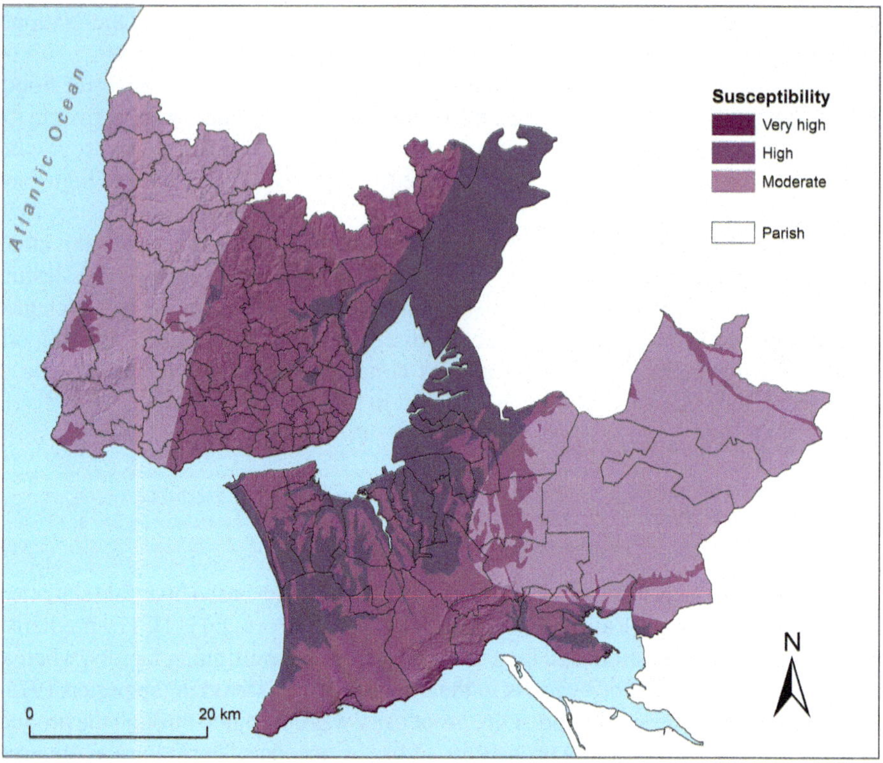

Fig. 2 Earthquake susceptibility in the Lisbon Metropolitan Area

intersections: (i) seismic intensity of degree IX or X + PGA from 3.2 to 4.0 m/s^2; (ii) seismic intensity of degree IX or X + PGA from 2.4 to 3.2 m/s^2 + existence of unconsolidated Quaternary deposits and/or active faults; (iii) seismic intensity of degree VIII + PGA of 3.2 to 4.0 m/s^2 + existence of unconsolidated Quaternary deposits and/or active faults. The moderate susceptibility class covers 41.4% of the LMA and corresponds to the following intersections: (i) seismic intensity of degree IX or X + PGA from 2.4 to 3.2 m/s^2; (ii) seismic intensity of degree VIII + PGA from 3.2 to 4.0 m/s^2; (iii) seismic intensity of degree VIII + PGA from 1.6 to 3.2 m/s^2.

3.2 Tsunami

The results of tsunami numerical modeling (Fig. 3) show that the tsunami inundates all beaches in the LMA. Additionally, coastal, riverine, and estuarine areas with elevations less than 3 m a.s.l. show inundation extent up to about 2.5 km. The coastal zone of Costa da Caparica, together with the inner zones of Tagus and Sado estuaries, are the most exposed to the tsunami hazard.

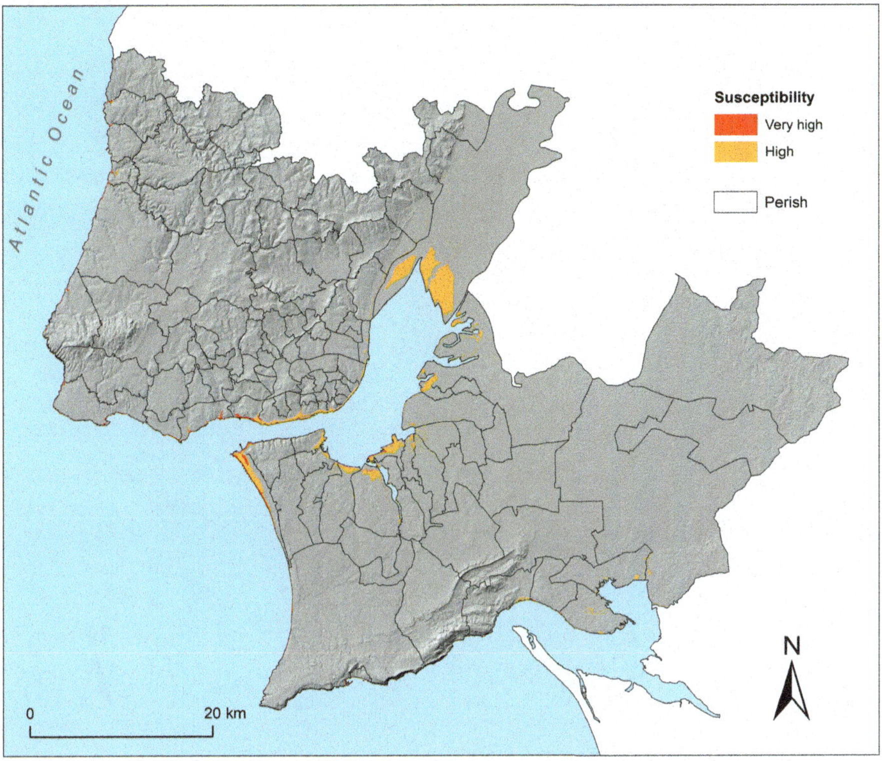

Fig. 3 Tsunami flooding susceptibility in the Lisbon Metropolitan Area

The very high class of tsunami flood susceptibility (water height above 2 m) covers 4.52 km^2, whereas the high susceptibility class (water height ranging from 0.1 to 2 m) covers 43.51 km^2.

3.3 Beach Erosion and Coastal Flooding

The beach erosion and coastal flooding susceptibility in the LMA are shown in Fig. 4. The very high susceptibility class corresponds to the setback line projected for 2050, whereas the high susceptibility class corresponds to the additional area between the previous line and the setback line projected for 2100. The coastal zone between the Meco beach and the Cova do Vapor is the most susceptible to beach erosion and coastal flooding within the LMA.

The very high susceptible zone extends over an area of 2.22 km^2, while the high susceptible zone covers an area of 3.75 km^2.

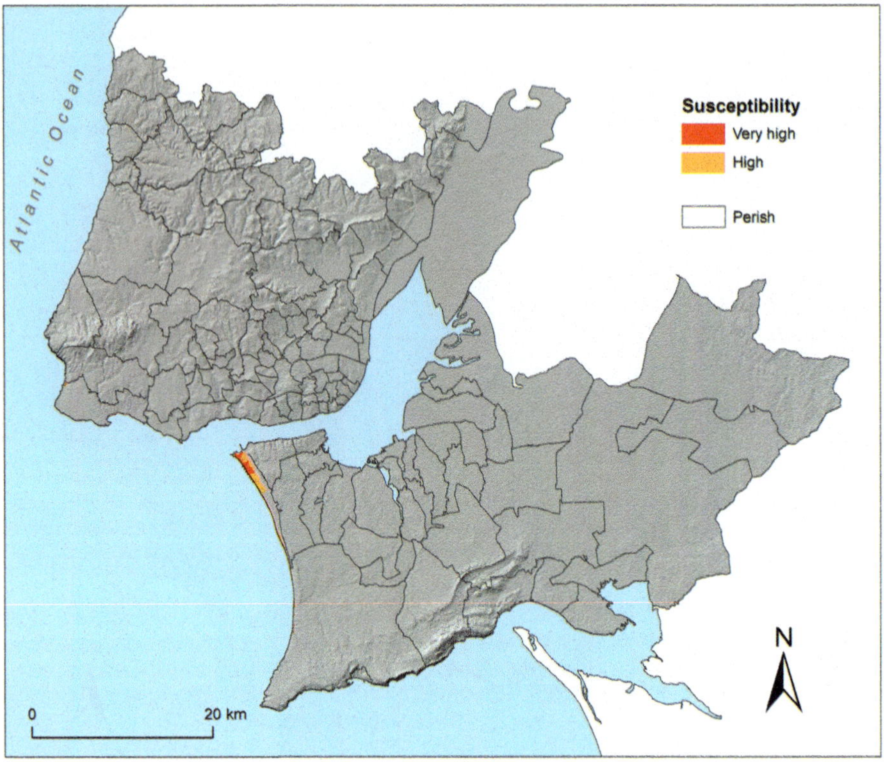

Fig. 4 Beach erosion and coastal flooding susceptibility in the Lisbon Metropolitan Area

3.4 Beach Erosion and Coastal Flooding

The coastal erosion and cliff retreat susceptibility in the LMA are represented in Fig. 5. Besides the identification of the unstable coastal slopes, two susceptibility zones are distinguished: very high and high. The first corresponds to the risky zone adjacent to the cliff crest, while the second reproduces the additional protection zone, as defined in the Coastal Zone Programmes. Figure 5 also represents the zone adjacent to the cliff bottom that is susceptible to the propagation of slope movements originating from the cliff.

The unstable coastal slopes cover an area of 4.74 km^2. The very high susceptible zone follows the cliff crest and extends over 7.13 km^2, while the high susceptible zone covers an area slightly smaller (5.92 km^2). The zone of potential impact adjacent to the cliff bottom covers 2.45 km^2 and extends to the sea floor in many areas, like on the Arrábida coast.

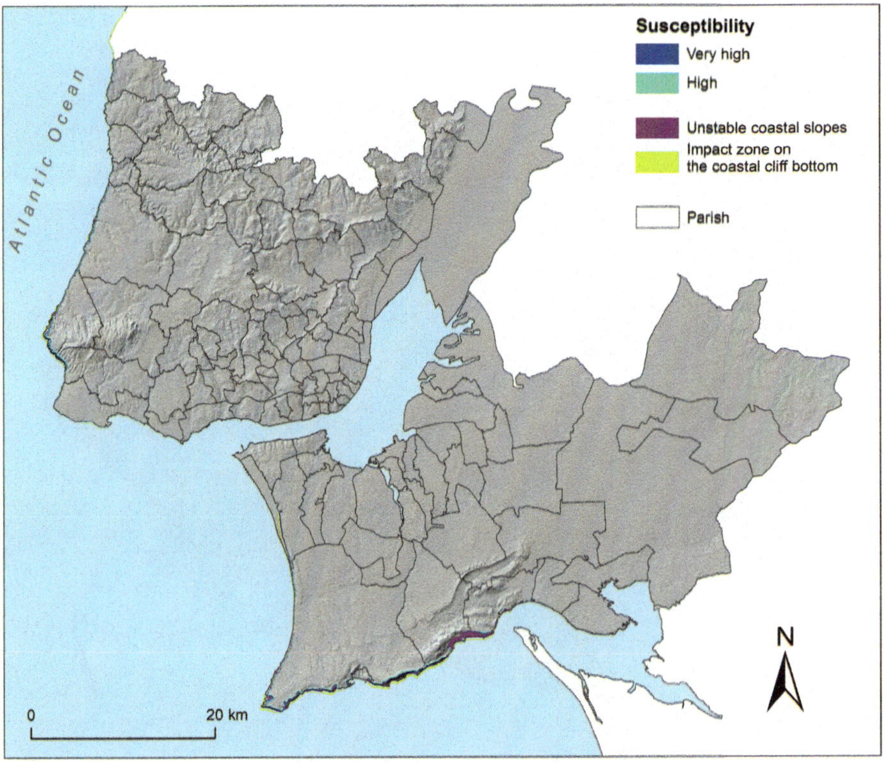

Fig. 5 Coastal erosion and cliff retreat susceptibility in the Lisbon Metropolitan Area

3.5 *Landslides*

Figure 6 illustrates the distribution of the landslide susceptibility in the LMA, which is highest along the Arrábida chain and in the hilly landscape located north of Lisbon. The very high susceptibility class validates 50% of the landslide inventory and covers 6.7% of the LMA, while the high susceptibility class validates 20% of the landslide inventory and extends over 4.9% of the study area. The moderate susceptibility class covers 7.8% of the LMA and the two lowest susceptibility classes cover, together, 80.6% of the total surface in the LMA.

3.6 *Floods*

The areas susceptible to flooding cover 444.78 km^2 in the LMA (Fig. 7), most of which correspond to estuarine flooding in the Tagus and Sado estuaries (179.14 km^2, 40.3%). The area susceptible to riverine flooding occurs along the lower Tagus valley

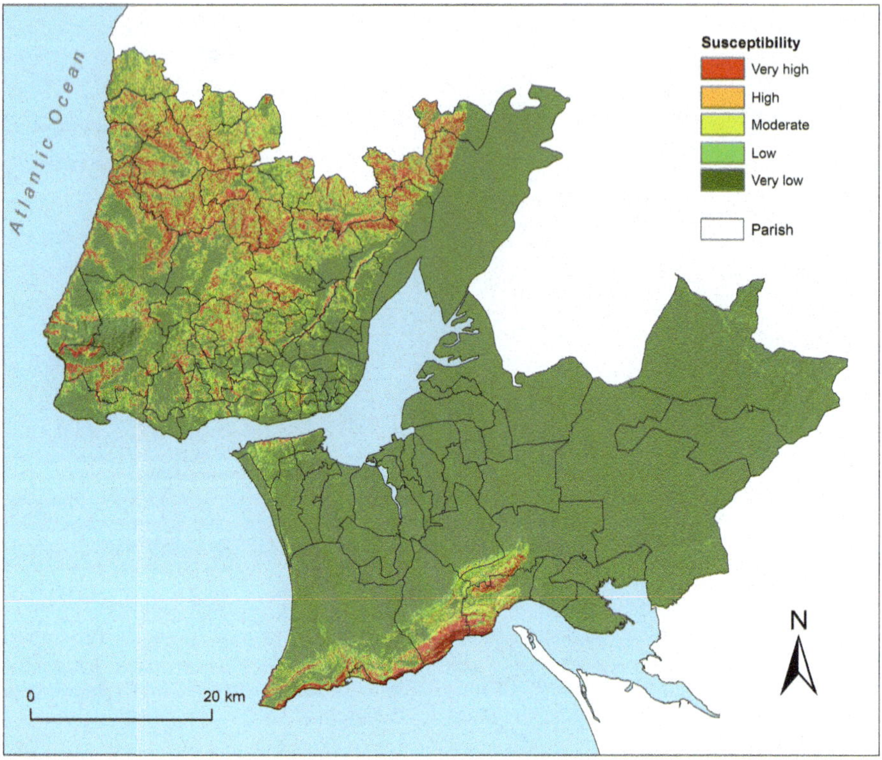

Fig. 6 Landslide susceptibility in the Lisbon Metropolitan Area

and corresponds to 32.5% (144.28 km^2) of the total flood susceptible area. The areas susceptible to flash flood cover 121.36 km^2 (27.3% of the total flood susceptible area).

3.7 Forest Fires

Figure 8 illustrates the distribution of forest fire susceptibility in the LMA. Artificialized areas, wetlands, and water bodies are not exposed to this hazard and were not classified. The forest fire susceptibility is highest along the Arrábida chain, the Sintra mountain, and the hilly zones of Mafra and Loures. The highest susceptibility classes (very high and high) have a small territorial expression (1.6 and 9.5% of the total classified area, respectively). The moderate susceptibility class covers 15.2% of the area, whereas the two lowest susceptibility classes cover together 73.7% of the total classified area.

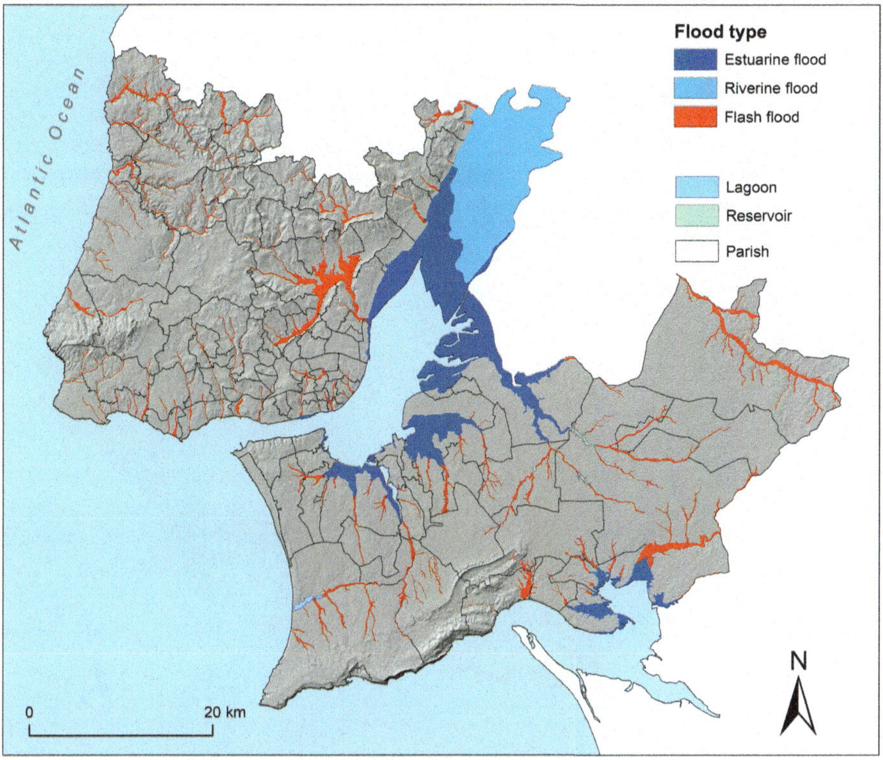

Fig. 7 Flood susceptibility in the Lisbon Metropolitan Area

4 Multi-hazard Susceptibility Assessment

Fig. 9 represents the multi-hazard susceptibility zoning of the Lisbon Metropolitan Area showing in superposition:

1. The three earthquake susceptibility zones that constitute the background of the map;
2. The areas subjected to tsunami flooding (susceptibility very high + high);
3. The areas subjected to beach erosion and coastal flooding (susceptibility very high + high);
4. The areas subjected to coastal erosion and cliff retreat (susceptibility very high + high + unstable coastal slopes + impact zones on the coastal cliff bottom);
5. The most susceptible areas to landslide occurrence (susceptibility very high + high);
6. The areas susceptible to floods (estuarine, riverine, and flash floods);
7. The most susceptible areas to forest fire occurrence (susceptibility very high + high).

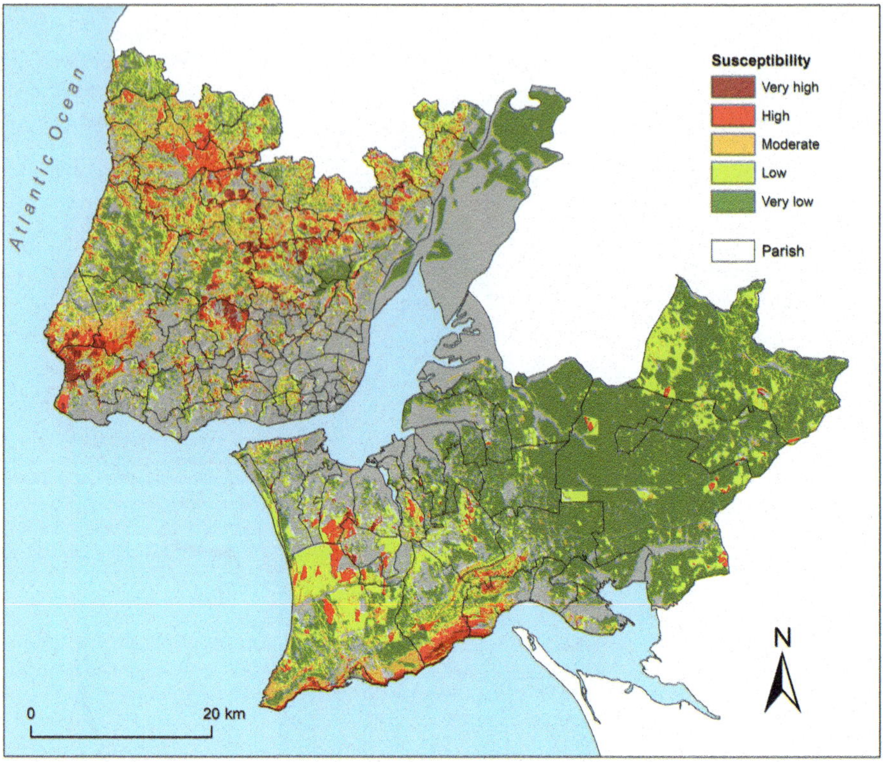

Fig. 8 Forest fire susceptibility in the Lisbon Metropolitan Area

The layers contained in the multi-hazard map were used to re-scale the susceptibility for the parish level into a scale ranging from 0 to 5, based on the territorial expression of each hazard and respecting the rules described in Table 1.

Table 2 summarizes the weighting of the considered seven hazard processes, based on the Analytic Hierarchy Process. The AHP was performed during a brainstorm involving all the authors of this chapter and considered the typical frequency-magnitude relationship of each process as well as the corresponding damage capacity.

Three processes stand out as the most important, namely for spatial planning purposes: earthquakes, floods, and beach erosion and coastal flooding. In the following position of the ranking, three other processes were identified: coastal erosion and cliff retreat, landslides, and forest fires. Tsunami has the lowest ranking, essentially due to its low probability of occurrence.

Figure 10 shows the multi-hazard susceptibility obtained for each parish by adding the susceptibility scores of Table 1 that were weighted with values of Table 2. The five susceptibility classes were defined using the standard deviation method.

Eleven parishes exhibit a very high multi-hazard index, having in common a high earthquake susceptibility score (4 or 5). The Costa da Caparica (parish ID: 19)

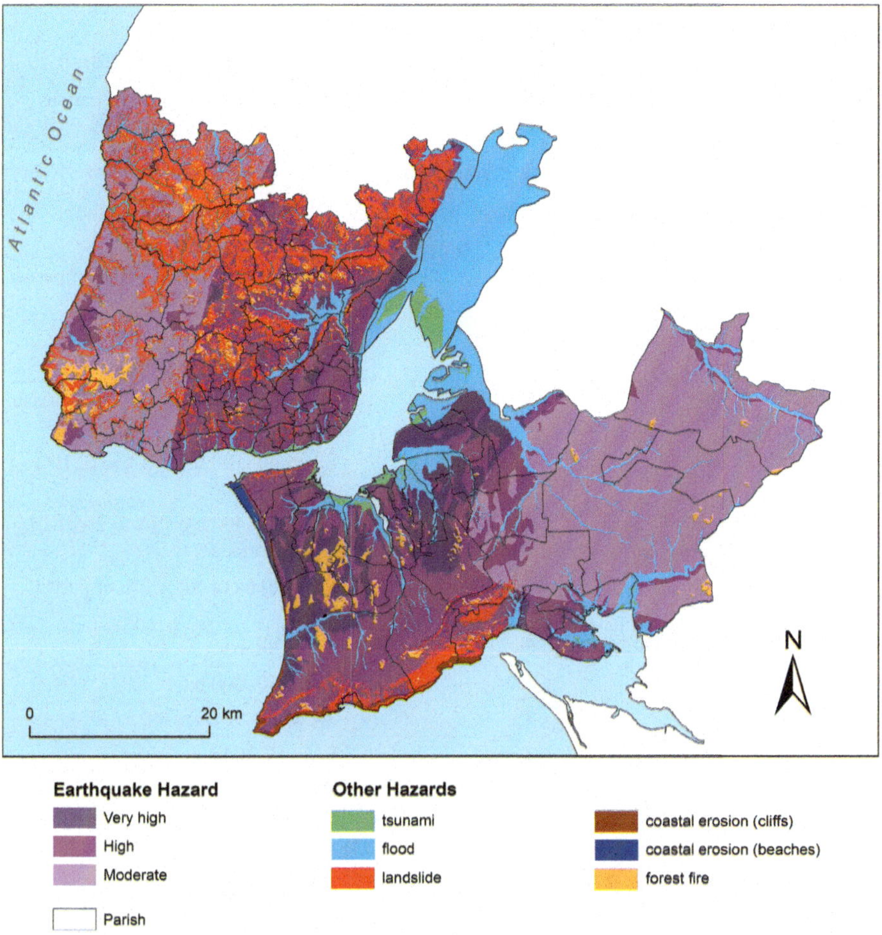

Fig. 9 Multi-hazard susceptibility in the Lisbon Metropolitan Area

registers the highest multi-hazard index, essentially due to the high susceptibility scores associated with earthquake, tsunami, beach erosion, and coastal flooding. The Arrábida territory, in the southern part of LMA, includes 4 parishes covered by the very high multi-hazard index class (Sesimbra-Castelo (ID: 6), Azeitão (ID:7), Setúbal (ID:3), and Sado (ID:2)), essentially due to the combination of the following hazards: earthquake, tsunami, flood, landslide and coastal erosion, and cliff retreat. Also, the Tagus estuary includes 4 parishes in this multi-hazard class (Samouco (ID:46), Alcochete (ID:86), Vila Franca de Xira (ID: 116), and Baixa da Banheira-Amoreira (ID: 24)), due to earthquake, tsunami, and floods. Finally, the very high multi-hazard index class covers 2 parishes north of the Tagus River: Cachoeiras (ID: 113) and S. António dos Cavaleiros-Frielas (ID: 89). The former is explained by the

Table 1 Rules to re-scale hazard susceptibility for the parish level

Hazard process	Susceptibility score	Rule (SA: Susceptible area)
Earthquakes	5	Very high class > 50% of parish surface or very high class > 35% and moderate class absent
	4	High class > 50% and very high class < 35% of parish surface, and moderate class is absent
	3	Moderate class = 50–90% of parish surface
	2	Moderate class > 90% of parish surface
Tsunami	5	SA > 100 ha or SA > 10% of parish surface
	4	SA = 40–100 ha or SA = 5–10% of parish surface
	3	SA = 20–40 ha or SA = 2–5% of parish surface
	2	SA < 20 ha or SA < 2% of parish surface
Beach erosion and coastal flooding	5	SA > 200 ha
	4	SA = 50–200 ha
	3	SA = 10–50 ha
	2	SA = 1–10 ha
	1	SA < 1 ha
Coastal erosion and cliff retreat	5	SA > 200 ha
	4	SA = 100–200 ha
	3	SA = 50–100 ha
	2	SA = 10–50 ha
	1	SA < 10 ha
Landslides	5	SA > 20% of parish surface
	4	SA = 10–20% of parish surface
	3	SA = 5–10% of parish surface
	2	SA = 0–5% of parish surface
	1	SA = 0% of parish surface
Floods	5	SA > 30% of parish surface
	4	SA = 20–30% of parish surface
	3	SA = 10–20% of parish surface
	2	SA = 5–10% of parish surface
	1	SA = 0–5% of parish surface
Forest fires	5	SA > 20% of parish surface

(continued)

Table 1 (continued)

Hazard process	Susceptibility score	Rule (SA: Susceptible area)
	4	SA = 10–20% of parish surface
	3	SA = 5–10% of parish surface
	2	SA = 0–5% of parish surface
	1	SA = 0% of parish surface

Table 2 Weighting of hazard processes considered in the Lisbon Metropolitan Area based on Analytic Hierarchy Process (AHP)

Hazard process	Weight
Earthquakes	0.234
Tsunamis	0.050
Beach erosion and coastal flooding	0.234
Coastal erosion and cliff retreat	0.083
Landslides	0.083
Floods	0.234
Forest fires	0.083

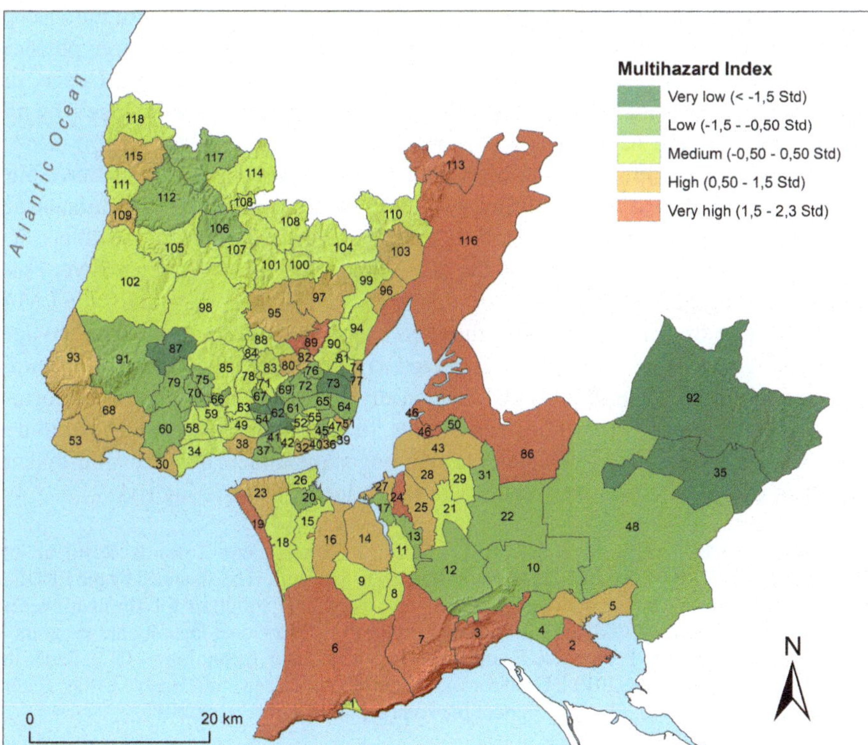

Fig. 10 Multi-hazard susceptibility in parishes belonging to the Lisbon Metropolitan Area. Numbers represent the parish ID

high susceptibility scores of floods and landslides, whereas the latter is explained by forest fire together with floods and landslides.

In opposite condition, 7 parishes exhibit a very low multi-hazard index, having in common the location far from the coast and the geographic position in the upper part of hydrographic basins, which explains the low susceptibility to floods: Algueirão-Mem Martins (ID: 87), Águas Livres (ID: 57), Benfica (ID: 62), Olivais (ID: 73), Moscavide and Portela (ID: 74), Canha (ID: 92), and Pegões (ID: 35).

5 Final Remarks

This work assessed the susceptibility to the occurrence of 7 natural and environmental hazards in the Lisbon Metropolitan Area. The susceptibility zonation maps were crossed with the 118 parishes that constitute the LMA, which were used as reference terrain units. Therefore, the susceptibility of occurrence of each hazard in each parish was re-classified on a scale of susceptibility scores (from 1 to 5), based on the territorial fraction of the parish exposed to that hazard.

The multi-hazard susceptibility of each parish resulted from the sum of the susceptibility scores, weighted based on an AHP, which valued essentially three physical processes: earthquakes, floods, and beach erosion and coastal flooding. Hence, the results of the multi-hazard susceptibility at the parish level is influenced by the decisions taken in the AHP.

All the parishes exhibiting very high multi-hazard susceptibility index have high earthquake susceptibility. In addition, with only two exceptions, these parishes are located in riverine or coastal zones, thus subjected to floods and/or coastal erosion (affecting beaches and/or cliffs). On the other hand, parishes with low and very low multi-hazard susceptibility index are typically located in the inner part of the LMA, in the upper part of small watersheds, thus less exposed to floods and coastal erosion.

The Lisbon Metropolitan Area is subjected to natural and environmental hazards, and some highly susceptible zones are densely inhabited and include important economic activities and critical infrastructures. In these circumstances, the obtained results should be considered by land use planning stakeholders and decision-makers in order to prevent the use of hazardous zones and mitigate existing risks.

Acknowledgements The project "MIT-RSC—Multi-risk Interactions Towards Resilient and Sustainable Cities" (MIT-EXPL/CS/0018/2019) leading to this work is co-financed by the ERDF—European Regional Development Fund through the Operational Program for Competitiveness and Internationalisation—COMPETE 2020, the North Portugal Regional Operational Program—NORTE 2020, and by the Portuguese Foundation for Science and Technology—GCT under the MIT Portugal Program at the 2019 PT call for Exploratory Proposals in "Sustainable Cities". Pedro P. Santos is financed through FCT I.P., under the contract CEECIND/00268/2017.

References

1. AML (2022) https://www.aml.pt/index.php. Accessed November 2022
2. Cabral J, Ribeiro A (1988) Carta neotectónica de Portugal continental 1: 1 000 000. Serviços Geológicos de Portugal
3. Carrilho F, Custódio S, Bezzeghoud M, Oliveira CS, Marreiros C, Vales D, Alves P, Pena A, Madureira G, Escuer M, Silveira G, Corela C, Matias L, Silva M, Veludo I, Dias NA, Loureiro A, Borges JF, Caldeira B, Wachilala P, Fontiela J (2021) The Portuguese national seismic network products and services. Seismol Res Lett 92:1541–1570
4. Carvalho J, Cabral J, Gonçalves R, Torres L, Mendes-Victor L (2006) Geophysical methods applied to fault characterization and earthquake potential assessment in the Lower Tagus Valley, Portugal. Tectonophysics 418:277–297
5. Grácia E, Donabeitia J, Vergés J, PARSIFAL Team (2003) Mapping active faults offshore Portugal (36°N–38°N): implications for seismic hazard assessment along the shouthweast Iberian margin. Geology 31(1):83–86
6. Guerreiro M, Fortunato AB, Freire P, Rilo A, Taborda R, Freitas MC, Andrade C, Silva T, Rodrigues M, Bentin X, Azevedo A (2015) Evolution of the hydrodynamics of the Tagus estuary (Portugal) in the 21st century. Revista de Gestão Costeira Integrada-J Integr Coast Zone Manag 15(1):65–80
7. Imamura F (1995) Review of tsunami simulation with a finite difference method. In: Long-wave runup models. World Scientific, Singapore, pp 25–42
8. Marques F, Andrade C, Taborda R, Freitas C, Antunes C, Mendes T, Carreira D (2009) Zonas costeiras. In: Santos FD (ed) Plano estratégico do concelho de sintra face às alterações climáticas, Câmara Municipal de Sintra, 62 pp
9. Marques F, Penacho N, Queiroz S, Gouveia L, Matildes R, Redweik P (2013) Estudo da adequabilidade das faixas de risco/salvaguarda definidas no POOC em vigor, Entregável 1.3.3.a, Estudo do litoral na área de intervenção da APA, I.P./ARH do Tejo, Agência Portuguesa do Ambiente
10. Okada Y (1985) Surface deformation due to shear and tensile faults in a half space. Bull Seismol Soc Am 75:1135–1154
11. Oliveira S, Gonçalves A, Zêzere JL (2021) Reassessing wildfire susceptibility and hazard for mainland Portugal. Sci Total Environ 762:143121
12. Peláez Montilla JA, López Casado C (2002) Seismic hazard estimate at the Iberian Peninsula. Pure Appl Geophys 159(11):2699–2713
13. Penacho N, Marques F, Queiroz S, Gouveia L, Matildes R, Redweik P, Garzón V (2013) Inventário de instabilidades nas arribas obtido por fotointerpretação, Entregável 1.2.2.1.a, Estudo do litoral na área de intervenção da APA, I.P./ARH do Tejo, Agência Portuguesa do Ambiente
14. Penacho N, Marques F, Queiroz S, Gouveia L, Matildes R, Redweik P, Garzón V (2013) Determinação e cartografia da perigosidade associada à ocorrência de fenómenos de instabilidade em arribas à escala regional, Entregável 1.3.1.a, Estudo do litoral na área de intervenção da APA, I.P./ARH do Tejo, Agência Portuguesa do Ambiente
15. Saaty TL (1988) What is the analytic hierarchy process? In: Mathematical models for decision support. Springer, Berlin, Heidelberg, pp 109–121
16. Saaty TL (1991) Some mathematical concepts of the analytic hierarchy process. Behaviormetrika 18(29):1–9
17. Santos A, Koshimura S, Imamura F (2009) The 1755 Lisbon Tsunami: Tsunami source determination and its validation. J Disaster Res 4(1):41–52
18. Silva AN, Taborda R, Lira C, Andrade CF, Silveira TM, Freitas MC (2013) Determinação e cartografia da perigosidade associada à erosão de praias e ao galgamento oceânico. Entregável 1.3.2.a, Estudo do litoral na área de intervenção da APA, I.P./ARH do Tejo, Agência Portuguesa do Ambiente
19. Silva NA, Taborda R, Lira C, Andrade CF, Silveira TM, Freitas MC (2013) Determinação e cartografia da perigosidade associada à erosão de praias e ao galgamento oceânico na Costa

da Caparica. Entregável 2.4.a, Estudo do litoral na área de intervenção da APA, I.P./ARH do Tejo, Agência Portuguesa do Ambiente
20. Taborda R, Andrade C, Marques F, Freitas M, Rodrigues R, Antunes C, Pólvora C (2010) Plano estratégico de Cascais face às alterações climáticas-Sector zonas costeiras
21. Trigo R, Ramos C, Pereira S, Ramos A, Zêzere JL (2016) The deadliest storm of the 20th century striking Portugal: flood impacts and atmospheric circulation. J Hydrol 541(A):597–610
22. Yin KL, Yan TZ (1988) Statistical prediction models for instability of metamorphosed rocks. In: International symposium on landslides, vol 5, pp 1269–1272
23. Zêzere JL (2002) Landslide susceptibility assessment considering landslide typology. A case study in the area north of Lisbon (Portugal). Natural Hazards and Earth System Sciences, vol 2, 1/2, pp 73–82
24. Zêzere JL (2020) Geomorphological hazards. In: Landscapes and landforms of Portugal. Springer, Cham, pp 47–62

The manufacturer's authorised representative in the EU is Springer Nature Customer Service Centre GmbH, Europaplatz 3, 69115 Heidelberg, Germany. If you have any concerns regarding our products, please contact ProductSafety@springernature.com

Printed and bound by CPI Group (UK) Ltd, Croydon, CR0 4YY
30/09/2026
02225308-0012